T0225280

# Advancing Food Integrity

## GMO Regulation, Agroecology, and Urban Agriculture

# Praise for the Book

*Advancing Food Integrity* provides a comprehensive and analytical overview of the contemporary challenges faced by government, industry, and civil society in an increasingly globalized world troubled with issues of climate change, urbanization, scientific advancement, and food security and safety. It offers not only a scholarly account to map such systematic, cutting-edge food integrity problems, but also optimal and innovative ways to solve them. This fascinating and timely book will be of great interest to researchers and practitioners of food law, environmental law, and agriculture and sustainability.

**Ching-Fu Lin**
*Assistant Professor of Law at National Tsing Hua University (NTHU)*

There are provocative and controversial ideas in this book, chief among them, the very concept of food integrity and the role of GMOs. Whether or not you agree, this book deserves your attention. The food system is inherently provocative, inherently controversial, because food, the environment, and human and animal well-being are at the same time essential and complex, evading easy answers at every turn. This book will expose you to perspectives that will help you navigate this intricate system.

**Joshua Ulan Galperin**
*Yale University*

JFK in a speech to Congress on March 15, 1962, spoke of consumers rights being central to the work of every agency and branch of the US government and that their voice was not heard loud enough. Gabriela Steir's pioneering book highlights all these years later that still in many areas of law, in particular in relation to food law, the consumers' voice is still not heard compared to the interests of trade and the market.

Gabriela covers the issues of food safety, food sovereignty, food security, environmental sustainability, and climate change in relation to GMOs from the perspective of a well-versed food lawyer, in order to demonstrate where private, public, and international are at fault.

Her approach gives new insights into the divide between the United States and the EU food regulatory regimes whether over hormone-raised beef, chlorinated chicken, or GMOs. These food regulatory issues were one of the main reasons why the EU-US TTIP negotiations failed last year.

The role of science is crucial here with the United States saying their food law is science driven, while the EU also says its law is backed-up by risk assessments from the European Food Safety Authority (EFSA). Yet, the division between the two systems is the famous *precautionary principle* utilized in EU law, which the United States abhors. Gabriela highlights that the *precautionary principle* is included within the Cartagena Protocol on Biosafety yet, surprise, surprise, the United States is not a signatory to this international agreement.

Gabriela's book is pertinent at this time in order to highlight the need for a more systematic look at why private, public, and international law are not adequately advancing food integrity for the benefit of consumers and the environment.

**Raymond O'Rourke, LLB, BL**
*Food & Consumer Lawyer, Ireland*

Gabriela Steier leads the reader on a journey beyond the boundaries separating private and public law, through an intriguing dialogue between legal, social, and life sciences.

As the book vividly explains, the emerging paradigm of agroecology can be turned into the pillar of a new legal architecture of food systems, in order to fulfill a crucial objective to the long-term survival of human beings on the planet: food integrity, as a legal synthesis of food safety, food sovereignty, food security, environmental sustainability, and climate change resilience.

This book is warmly recommended to all those readers who believe in the possibility of building hope for the future by reconciling the "ought to be" of agrifood law with the "to be" of food and environmental sciences.

**Prof. Dr. Massimo Monteduro**
*Associate Professor of Administrative and Environmental Law,*
*University of Salento*

In *Advancing Food Integrity*, Gabriela Steier looks critically at the performance of regulatory schemes for agriculture and food systems, and their potential to promote food integrity. The analysis, covering both private and public regulation at the national and international scale, is conceptually, empirically, and methodologically astute. The comprehensive, updated, multidisciplinary perspective offers an essential contribution to the scientific understanding of food policies. The book should be required reading for students of environmental law and policy, agricultural studies, and political economy.

**Ronit Justo-Hanani**
*Tel Aviv University*

# Advancing Food Integrity
## GMO Regulation, Agroecology, and Urban Agriculture

by
Gabriela Steier

CRC Press
Taylor & Francis Group
Boca Raton London New York

CRC Press is an imprint of the
Taylor & Francis Group, an **informa** business

CRC Press
Taylor & Francis Group
6000 Broken Sound Parkway NW, Suite 300
Boca Raton, FL 33487-2742

First issued in paperback 2020

© 2018 by Taylor & Francis Group, LLC
CRC Press is an imprint of Taylor & Francis Group, an Informa business

No claim to original U.S. Government works

ISBN 13: 978-0-367-57252-5 (pbk)
ISBN 13: 978-1-138-30525-0 (hbk)

This book contains information obtained from authentic and highly regarded sources. Reasonable efforts have been made to publish reliable data and information, but the author and publisher cannot assume responsibility for the validity of all materials or the consequences of their use. The authors and publishers have attempted to trace the copyright holders of all material reproduced in this publication and apologize to copyright holders if permission to publish in this form has not been obtained. If any copyright material has not been acknowledged, please write and let us know so we may rectify in any future reprint.

Except as permitted under U.S. Copyright Law, no part of this book may be reprinted, reproduced, transmitted, or utilized in any form by any electronic, mechanical, or other means, now known or hereafter invented, including photocopying, microfilming, and recording, or in any information storage or retrieval system, without written permission from the publishers.

For permission to photocopy or use material electronically from this work, please access www.copyright.com (http://www.copyright.com/) or contact the Copyright Clearance Center, Inc. (CCC), 222 Rosewood Drive, Danvers, MA 01923, 978-750-8400. CCC is a not-for-profit organization that provides licenses and registration for a variety of users. For organizations that have been granted a photocopy license by the CCC, a separate system of payment has been arranged.

**Trademark Notice:** Product or corporate names may be trademarks or registered trademarks, and are used only for identification and explanation without intent to infringe.

**Visit the Taylor & Francis Web site at**
**http://www.taylorandfrancis.com**

**and the CRC Press Web site at**
**http://www.crcpress.com**

# Dedication

---

*For Oma Fany*

# Contents

# Preface

Many years ago, my esteemed Italian professor at Tufts University, Isabella Perricone, taught me an important lesson about luxuries by telling me the story of Gianni Rodari Professor Grammaticus. This is how I remember it:

Once upon a time, Professor Grammaticus was traveling by train in his native Italy. Two seasonal workers joined his compartment at a stopover and he started a conversation with them. As they told him where they had started their journey of harvesting this year's crops, he could not help but correct their grammatical errors. Time and again, as they were recounting their fate, he interrupted them, and they thanked him with utmost respect and continued to tell the heartbreaking stories of their rough lives and unreliable labor.

When Professor Grammaticus began to teach about transitive and intransitive verbs, one of the workers said to him that he felt that "going to work" was an undoubtedly important one but a very sad one. He said that going to work at other people's houses means leaving one's family behind for months at a time to earn money. Professor Grammaticus began to stammer and continued his explanations of grammatical intricacies. Again, the worker interrupted him gently and smiled heavy-heartedly, "I am, we are, they are ... but where we are, with all of the verb *to be* and all of our hearts is always at home. We *have been* to Germany and Belgium, *we are* always here and there, but where we really want *to be* is to be home." The workers explained that they had to travel to send money back home for their wives and families, and they rarely had a comfortable place to rest and think about grammar. They stopped and looked at the professor with their tired and sad eyes. At this point, Professor Grammaticus wanted to throw both of his hands over his head. He realized that he was looking for grammatical errors, while *his* gravest errors were elsewhere.

Just like Professor Grammaticus, I often feel that writing about food is a luxury. I have the luxury to think about the food system on a full stomach. In fact, indulging in difficult theoretical questions of international comparative food law is a luxury in and of itself, and I am grateful that I have the opportunity to indulge. It is, to a large extent, thanks to my parents and grandparents' unwavering support and my son and husband's patience that I can afford this luxury. I am a little like Professor Grammaticus, watching farm workers from the train of life.

However, I also consider scholarship as a tool to provide guidance to the real and everyday problems that those too close to the problems cannot see or even comprehend. The abstractions afforded in my research may, therefore, prove useful when policy makers, regulators, advocates, grassroots organizations, or others reach an impasse. I humbly hope that my work provides vision and integrity that inspires others not to give up control of our food system, of the choices about how to nourish our bodies, and of how to protect our planet. After all, what we can access in order to nourish our bodies also affects our minds, families, and, as this book shows, the economy and even the international market. Thus, if I were in Professor Grammaticus' shoes, I would pledge to the workers to hold down the fort of the world of abstractions and be grateful for their contribution to the whole, for we all play important parts from farm to fork and in between to make ours a better and safer world.

I am completing this book shortly after earning the last few credits for my LLM in Food and Agriculture Law from the Vermont Law School. The work I have done as an LLM student has inspired and provided the basis for parts of this book. This book, however, represents so much more than merely a compilation of research. It is, as the German turn of phrase would describe it, the *Herzblut*, that is, the blood, sweat, and tears of my work as a food lawyer, a mother, a daughter, a wife, and a granddaughter. This book is my contribution toward a better world—my hope for an interdisciplinary study and collaboration on an international scale to revolutionize how we obtain our food, how we maintain the integrity of our nourishment and its production, and how we protect this beautiful world so that our children and children's children can enjoy it. Hopefully, this book will inspire scholars, lawyers, scientists, policy analysts, and stakeholders of any caliber to take action, to question the status quo, and to turn the food system over toward food integrity.

**Gabriela Steier**
*Boston, MA*

# Acknowledgments

First, I am grateful to Professor Dr. Kirk W. Junker from the University of Cologne for giving me the opportunity to pursue a doctorate in comparative law under his exceptional guidance. It has been an unforgettable privilege and an honor to have him as my *Doktorvater* (thesis advisor). This book is the fruit of my research, and I hope that it will make him proud.

Additionally, I am grateful to the professors I had during my LLM studies at the Vermont Law School. Many of the topics and much of the research included in this book had their start in my studies there.

Most of all, I am grateful to my parents and my grandmother for their unwavering support, encouragement, and love, and to my husband and son for their patience and endurance that made indulging in the work for this book so much sweeter.

A special note of gratitude goes to Professor Bernard Rollin, Professor Alberto G. Cianci, and John Sulzycki for reviewing and championing this book, which will be published so it can reach a wide audience.

# About the author

**Gabriela Steier, BA, JD, LLM** is a founding partner, Food Law International, www.foodlawinternational.com; visiting professor, University of Perugia, Department of Political Science, Perugia, Italy; an adjunct professor, Duquesne Law School, Pittsburgh, Pennsylvania, USA.

Gabriela Steier is co-founder of Food Law International (FLI) and editor-in-chief of the textbooks *International Food Law and Policy* (Springer, 2017) and *International Farm Animal, Wildlife and Food Safety Law* (Springer, 2017). She is an attorney and focuses on food safety, policy, animal welfare, and GMO issues domestically and in the European Union.

She worked as an LLM Fellow in Food and Agriculture Law at the Vermont Law School. She also joined the Duquesne University School of Law as an adjunct professor teaching a breakthrough new course in "Food Law and Policy" in 2015 and "Climate Change Law" in 2016. As visiting professor at the University of Perugia, Italy, she also teaches EU–US comparative food law at the Department of Political Sciences. She holds an LLM in Food and Agriculture Law from the Vermont Law School, a BA from Tufts University, a JD from Duquesne University, and a doctorate in comparative law from the University of Cologne in Germany. She worked as a legal fellow at the Center for Food Safety on Capitol Hill in Washington, DC, from 2013 until 2015.

Gabriela has published widely on international food law, policy, and trade and has earned several awards for her work. As an experienced editor and with her numerous publications ranging from peer-reviewed articles in international medical journals to law reviews, Gabriela has gained widespread interdisciplinary interest. Some of her articles have been on the top ten list on SSRN for several months, as well as the top ten list in environmental law textbooks on Amazon.com.

She lives with her family in Boston, Massachusetts, and speaks six languages. In her spare time, she enjoys painting, drawing, and nature walks.

# List of abbreviations

| | |
|---|---|
| **CJEU** | Court of Justice of the European Union |
| **DHHS** | United States Department of Health and Human Services |
| **EEA** | European Economic Area |
| **EFSA** | European Food Safety Authority |
| **EPA** | Environmental Protection Agency |
| **ESA** | Endangered Species Act |
| **EU** | European Union |
| **FAO** | Food and Agriculture Organization of the United Nations |
| **FDA** | United States Food and Drug Administration |
| **FIFRA** | Federal Insecticide, Fungicide, and Rodenticide Act |
| **FSMA** | Food Safety Modernization Act |
| **GE** | Genetic Engineering *or* Genetically Engineered |
| **GM** | Genetic Modification *or* Genetically Modified |
| **GMO** | Genetically Modified Organism |
| **GRAS** | Generally Recognized as Safe |
| **IPPC** | International Plant Protection Convention |
| **NEPA** | National Environmental Policy Act |
| **TFEU** | Treaty on the Functioning of the European Union |
| **TTIP** | Transatlantic Trade and Investment Partnership |
| **UN** | United Nations |
| **US** | United States of America |
| **USDA** | United States Department of Agriculture |
| **WHO** | World Health Organization |

*chapter one*

# Food integrity and the food system defined

Imagine a single chess player. In a game against himself, he is challenging his own strategies and viewpoints. He needs to switch the board around to plot the next move in the game against himself. Only a new perspective can help him find the right strategy in his mental tool box to tackle a seemingly "stuck" problem. Just like the chess player, the race toward improved food integrity is one against climate change and urbanization. These challenges may appear to be self-defeating and at an impasse when it comes to certain agricultural products, but changing the perspective helps to reevaluate problems and reveal paths to new solutions. This book is about finding perspective shifts to get our food system "unstuck." Here, the disciplinary lens is that of the law. Focusing on the food system, this book explores whether private law has sufficiently protected food or whether public law control is needed to safeguard food integrity.

First, it must be clear what this food system encompasses. Professors Robert Lawrence and Roni Neff from Johns Hopkins University define the food system as "encompassing all the activities and resources that go into producing, distributing, and consuming food; the drivers and outcomes of those processes; and all the relationships and feedback loops between system components."[1] Notably, the components of a food system include "land-based parts (e.g., agriculture, farmland preservation); environment (e.g., water, soil, energy); economy (e.g., distribution, processing, retail); education; policy; social justice; health; and food cultures."[2] How the relationships between these functional parts play out provide points of attack for "the most strategic and practical ways to intervene for change,"[3] such as through the perspective switchboard described herein.

## 1.1 Functional components of food integrity

Sharing the functional components, food integrity describes an ideal direction for the relationships and "shall be defined as the measure of

---

[1] Roni Neff (Ed.), *Introduction to the US Food System*. Wiley & Sons (2014); ProQuest Ebook Central.

[2] Id. Internal citations omitted.

[3] Id.

*1*

environmental sustainability[4] and climate change resilience, combined with food safety, security, and sovereignty for the farm-to-fork production and distribution of any food product."[5] Conversely, food safety, as the Food and Agriculture Organization (FAO) of the United Nations defines it, focuses on "handling, storing and preparing food to prevent infection and help to make sure that … food keeps enough nutrients for … a healthy diet."[6] Here, the key is the wholesome nature of food so that it nourishes instead of sickens. Food security, however, exists when "availability and adequate access at all times to sufficient, safe, nutritious food to maintain a healthy and active life"[7] are guaranteed. Three pillars of food security, (1) availability, (2) access, and (3) utilization, are designed to ensure that food is "available in sufficient quantities and on a consistent basis," that it is regularly acquirable at adequate quantities, and that it has "a positive nutritional impact on people."[8] Finally, food sovereignty "is the right of peoples to healthy and culturally appropriate food produced through ecologically sound and sustainable methods, and their right to define their own food and agriculture systems."[9] Notably, food sovereignty "puts those who produce, distribute and consume food at the heart of food systems and policies rather than the demands of markets and corporations."[10] In other words, food sovereignty creates a right that complements both food safety (the wholesomeness of food) and food security (the access to food). Adding the elements of environmental sustainability and climate change resilience, one speaks about food integrity (see Figure 1.1).

Second, the food system is maintained by a set of laws, ranging from local ordinances about animal husbandry to domestic farming and international trade. Without going into the specifics yet, this book uses genetically modified organisms (GMOs) as the most commonly traded commodity crops to illustrate how laws interact with the principles underlying food integrity.[11] Simply put, this book explores selected points

---

[4] For the purpose of this book, sustainability shall be defined as pertaining to a food system that maintains its own viability by using agroecologic techniques that allow for continual reuse and a wholistic service to all components of food integrity. Agricultural sustainability shall be construed to complement environmental conservation and climate change mitigation.

[5] Gabriela Steier, A window of opportunity for GMO regulation: Achieving food integrity through cap-and-trade models from climate policy for GMO regulation, 34 *Pace Envtl L. Rev.* 293 (2017).

[6] FAO, Food safety, http://www.fao.org/docrep/008/a0104e/a0104e08.htm.

[7] WFP. What is food security?, https://www.wfp.org/node/359289.

[8] Id.

[9] La Via Campesina, The declaration of Nyéléni, https://nyeleni.org/spip.php?page=NWarticle. en&id_article=479.

[10] Id.

[11] USDA-ERS, Corn and other feed grains: Trade, https://www.ers.usda.gov/topics/crops/corn/trade.aspx (last accessed May 29, 2017).

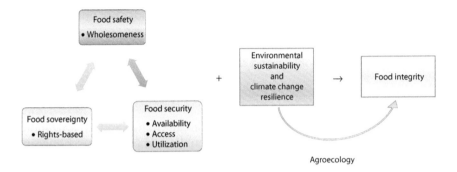

***Figure 1.1*** Food integrity. The combination of food safety (blue), food sovereignty (light green), and food security (dark green) with environmental sustainability and climate change resilience (orange) creates food integrity. Agroecology is one link (yellow arrow) that helps to achieve food integrity (yellow).

where the law obstructs achieving food integrity and suggests how the food system may be unclogged. For instance, private law facilitates GMO proliferation by creating an enabling environment for their production and trade. The record of private law and legal actions, as recorded in the many cases and court decisions cited in this book, shows that public law provides little possibility for critique or control in a systematic fashion that permeates all parts of the food system.[12] Instead, public law advances the interest of the market, prioritizing the trade and business aspects of the food system under the pretense of control by the market. This food trade in a free market context, however, plays a counterproductive role in achieving food integrity when one considers the GMO sector. Most of the requirements to maintain food security, food safety, and food sovereignty fall prey to economic drivers.[13] Conversely, public law could control the product, as in the case of plastics, but neoliberal economics of food products reliant on GMOs, justified by the industry's advertising claim that GMOs will "feed the world," prevents public law control from being implemented across the food system. Simply put, private law is blocking the path for public law to achieve food integrity.[14] Especially in light of the growing challenges of climate change, agriculture is at a crossroads. On the one hand, agriculture contributes to global warming and climate

---

[12] Summary based on e-mail correspondence and conversations with Professor Dr. Kirk Junker (on file with the author).

[13] See generally, Andrew Kimbrell, *The Fatal Harvest Reader.* Island Press (2002).

[14] Summary based on e-mail correspondence and conversations with Professor Dr. Kirk Junker (on file with the author).

change; on the other hand, it has the potential to adapt to and mitigate it.[15] At this junction, the clash between (1) agroecology, (2) food integrity, and (3) private law is the origin of the GMO debate that is commonly discussed in the media, politics, and at conferences.[16]

## 1.2   Summary

Improving food integrity in light of the climate change and urbanization challenges gives rise to clashes between public and private law, between GMOs and agroecology. These challenges may appear to be self-defeating and at an impasse when it comes to GMOs. This chapter is about finding perspective shifts to get our food system "unstuck." Focusing on the food system, this chapter explores whether private law has sufficiently protected food or whether public law control is needed to safeguard food integrity. Specifically, food integrity describes an ideal direction for these relationships and shall be defined as the measure of environmental sustainability and climate change resilience, combined with food safety, security, and sovereignty for the farm-to-fork production and distribution of any food product.

---

[15] Smith, P., Martino, D., Cai, Z., Gwary, D., Janzen, H., Kumar, P., McCarl, B., Ogle, S., O'Mara, F., Rice, C., Scholes, B., and Sirotenko, O. Chapter 8: Agriculture. In *Climate Change 2007: Mitigation. Contribution of Working Group III to the Fourth Assessment Report of the Intergovernmental Panel on Climate Change* (B. Metz, O.R. Davidson, P.R. Bosch, R. Dave, and L.A. Meyer, Eds.). Cambridge University Press (2007).

[16] For summaries of common arguments in the hotly contested GMO debate, see McKay Jenkins, *Food Fight: GMOs and the Future of the American Diet*. Penguin Random House (2017).

# chapter two

# The perspective switchboard: GMOs versus agroecology

In international comparative food law, the debate about genetically modified organisms (GMOs) versus agroecology, that is, the ecology of food systems,[1] is a question of perspective. How shifting this perspective affects the legal and policy analysis of the widespread prevalence of GMOs is the subject of this book. The following chapters show how the proliferation of GMOs[2] creates food insecurity, by denying people's access to food through food system centralization, and how industrial agricultural policy to-date largely trivializes the dangers of GMO monocultures to crop diversity and biodiversity, thereby weakening food systems. All of these negative tendencies are made possible through private law and the protection it awards to those controlling the food system. As outlined in this chapter, the subsequent legal analysis will identify points of attack to improve the current food system using public law as a tool, that is, to pierce the shroud of private legal protection that creates the GMO-centered enabling system. On its flip side, the legal analysis will also highlight where the effectiveness of legal tools ends and opportunities for solutions to problems of food insecurity, trade imbalances, and other challenges begin. The following analysis uses a perspective switchboard, where a shift in viewpoint uses a comparative legal examination of the facts.

Changing the perspective to evaluate the debate from a different angle has the promising effect of revealing new approaches and solutions to an old issue: centralization. How to reveal these new angles and leverage them against each other in the quest for a new perspective is a matter of comparative law, the subject of Chapter 2. From the proliferation of GMOs around the globe, it follows that the centralization of food systems goes hand in hand with the standardization of food production. The effective centralization of society, which started with urbanization

---

[1] C. Francis, G. Lieblein, S. Gliessman, T. A. Breland, N. Creamer, R. Harwood, L. Salomonsson, J. Helenius, D. Rickerl, R. Salvador, M. Wiedenhoeft, S. Simmons, P. Allen, M. Altieri, C. Flora, and R. Poincelot, Agroecology: The ecology of food systems, *Journal of Sustainable Agriculture* 22: 3 (2003), http://www.tandfonline.com/doi/abs/10.1300/J064v22n03_10 (last accessed May 29, 2017).

[2] GMO Compass, Cultivation of GM plants: Increase worldwide, no great change in Europe, http://www.gmo-compass.org (last accessed Mar. 8, 2017).

in the polis in Ancient Rome and is peaking in mega-metropolitan areas, such as New York City today, exemplifies this trend. From a food law perspective,[3] urbanization led to the outsourcing of agriculture, which fundamentally changed society and continues to do so in the developing world. This centralization, in turn, leads to standardization, far-reaching trade, and transnational networks in the food system with mutually affecting variables. Shifting the perspective on this outsourcing of food production away from metropolitan areas becomes a question of access to food for consumers, that is, food security; for farmers, that is, food sovereignty; and to regulating food, that is, food safety. Chapters 3 through 5 detail agrobiodiversity and GMO regulation for context in relation to all of these challenges.

With an inquiry into food security, sovereignty, and security in the context of urbanization, the selected case studies and examples position the strains of food trade in the free market as a type of perspective switchboard because the current food system relies on the trade of food. Some of the switches are laws, or the lack thereof, as this book reveals. Those with the most control and power turn the switches on the board, while others adapt to a path of least resistance to navigate the centralized food system and its trade network, as Chapter 6 shows. The navigable path that this book focuses on consists of a set of private and public laws that shape the food system. Here, the standardization of food production is the key to the switchboard because only through standardization in both law and production can food systems become as big as they are and dominate international trade. Again, shifting the perspective, standardization is possible through the centralization of genetic material, hence the proliferation of GMOs, which links the ideas throughout this book. Similarly, laws usually apply where large ideas become reality, another form of standardization by commonality, which, in turn, trivializes differences. Picking up and analyzing the legal context of the trivialization of various risks for food safety that create a permissible relationship and facilitate the proliferation constitute the backbone of this book and provide points of orientation throughout the perspective shifts and analysis that follow.

---

[3] Food law must be distinguished from agriculture law. According to Prof. Michael Roberts from the UCLA School of Law, "agricultural law, which is rooted in the concept of agrarianism or, agricultural exceptionalism," grants special legal exceptions for agricultural practices. In contrast to food law, which concerns itself with the facets outlined in Figure 2.1, "[a]gricultural law has spawned law practices, largely in rural areas, that represent farms as well as agricultural enterprises, such as seed and chemical input companies." Roberts, Michael T., *Food Law in the United States* (pp. 6–7). Cambridge University Press (2016). Kindle edition.

## 2.1    The basics: Links between GMOs, agroecology, and urban agriculture

The multiplicity of issues and cases mentioned in this book may seem random without an understanding of the links between GMOs, urban agriculture, and agroecology and how all of these relate to the overall argument that the proliferation of GMOs trivializes their dangers. Simply put, where do GMOs, urban agriculture, and agroecology belong on the switchboard? This section clarifies how the links create a ripple effect of GMO proliferation on the entire food system, from farm to fork, and provides context for urban agriculture and agroecology.

Figure 2.1 illustrates how the proliferation of GMOs, facilitated by plant patenting, limits farmers' access to commodity crop seeds, thereby disallowing seed saving and displacing agrobiodiversity with just a few staple GMO commodity crops. Conversely, the growth of the GMO market creates powers that lead to price volatility and control international food trade. The underlying centralization of the seed industry, BigAg's backbone, fueled the centralization of the food supply, which, in turn, further tightened BigAg's grasp on the seed industry, thereby stifling agrobiodiversity. This power allowed the industry lobby to exert political pressure and create economic distortions to further standardize the commodity crop market through GMOs and to trivialize the risks of doing so. The side effects and simultaneous facilitators, industrialization and urbanization, connect seamlessly to the externalities of the food system, pollution, and climate change. Through the continued augmentation and reduction of regulators and market barriers, that is, by circling around, the proliferation of GMOs moves this cycle. Each circle, however, is a point of attack, with the potential to disrupt this cycle and make room for solutions. Starting with the proliferation of GMOs as a fact (see top circle in Figure 2.1), one must focus on the patenting of plants as a cornerstone of the spread of GMOs. The United States, for instance, grows over 40% of the world's corn.[4] Corn is a GMO commodity crop used for human food and animal feed, planted on over 85 million acres of monoculture farm land[5] in the United States alone. In what has become known as the "wheat belt" of the United States, the adoption of GMOs has multiplied fourfold, with mainly wheat, soybeans, cotton, and corn being planted.[6]

---

[4] USDA-ERS, Recent trends in GE adoption, https://www.ers.usda.gov/data-products/adoption-of-genetically-engineered-crops-in-the-us/recent-trends-in-ge-adoption.aspx (last accessed Mar. 8, 2017).

[5] Mary Angelo, Jason Czarnezki, and Bill Eubanks, *Food, Agriculture, and Environmental Law* at 114 (Environmental Law Institute) Environmental Law Institute. Kindle edition.

[6] USDA-ERS, Recent trends in GE adoption, https://www.ers.usda.gov/data-products/adoption-of-genetically-engineered-crops-in-the-us/recent-trends-in-ge-adoption.aspx (last accessed Mar. 8, 2017).

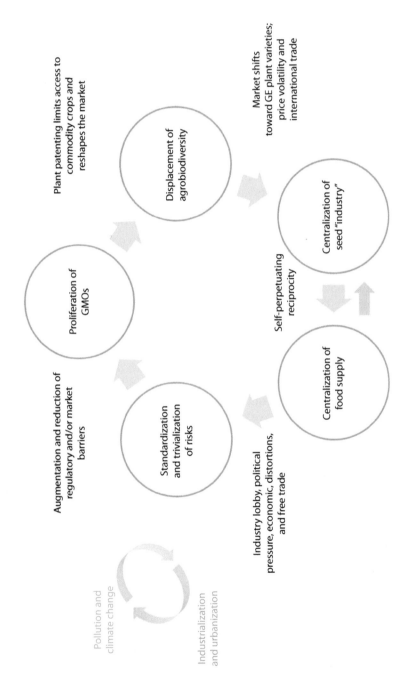

*Figure 2.1* The cycle linking GMOs, urban agriculture, and climate change.

The legal tools that make this GMO commodity crop dominance of the American farm-landscape possible are the US Plant Patent Protection Act (PPPA) of 1930[7] and the Plant Variety Protection Act (PVPA) of 1970,[8] which allow seed companies to patent plants. Additionally, in *Diamond v. Chakrabarty*,[9] the US Supreme Court ruled live organisms to be patentable. Coupled with *Pioneer Hi-Bred Int'l Inc. v. J.E.M. Ag Supply Inc. (Pioneer)*,[10] even utility patents[11] on GMOs are possible. As a result of the patent-ability of plants, especially of commodity agricultural crops, the agricultural industry (hereinafter referred to as BigAg) has blossomed,[12] and the proliferation of GMOs has been sparked. It has, in other words, taken over the US-American rural landscape. These private laws illustrate the previously noted enabling environment of the GMO proliferation.

Since plant patenting reshaped the landscape, it also affected the market, making BigAg bigger and more powerful. This, in turn, was realized in part by virtue of the benefits afforded by the 17 farm bills,[13] which put agriculture on a special pedestal in the United States. In fact, scholars name this pedestal and consider it "agricultural exceptionalism,"[14] which even awards agriculture its own executive agency, the US Department of Agriculture (USDA). Agricultural exceptionalism places BigAg somewhat above the law because it means that food production is often out

---

[7] 35 U.S.C. 161–164.

[8] 7 U.S.C. 2321 et seq.

[9] Diamond v. Chakrabarty, 447 U.S. 303, 100 S. Ct. 2204, 65 L. Ed. 2d 144 (1980).

[10] Pioneer Hi-Bred Int'l Inc. v. J.E.M. Ag Supply Inc., No. C 98-4016-DEO, 1998 WL 1120829 (N.D. Iowa Aug. 18, 1998), aff'd, 200 F.3d 1374 (Fed. Cir. 2000), aff'd, 534 U.S. 124, 122 S. Ct. 593, 151 L. Ed. 2d 508 (2001).

[11] 35 U.S.C. 101.

[12] Intellectual Property and Plants. Princeton.edu, https://www.princeton.edu/~ota/disk1/1989/8924/892407.PDF (last accessed Mar. 8, 2017).

[13] Renée Johnson and Jim Monke, Cong. Research Serv., RS22131, What is the farm bill?, at 1 n.1 (2014) (listing 17 farm bills since the 1930s: 2014, 2008, 2002, 1996, 1990, 1985, 1981, 1977, 1973, 1970, 1965, 1956, 1954, 1949, 1948, 1938, and 1933); *see also* Agricultural Adjustment Act of 1933, Pub. L. No. 73-10, 48 Stat. 31; Agricultural Adjustment Act of 1938, Pub. L. No. 75-430, 52 Stat. 31; Agricultural Act of 1948, Pub. L. No. 80-897, 62 Stat. 1247; Agricultural Act of 1949, Pub. L. No. 81-439, 63 Stat. 1051; Agricultural Act of 1954, Pub. L. No. 83-690, 68 Stat. 897; Agriculture Act of 1956, Pub. L. No. 84-540, 70 Stat. 188; Food and Agricultural Act, Pub. L. No. 89-321, 79 Stat. 1187; Agricultural Act of 1970, Pub. L. No. 91-524, 84 Stat. 1358; Agricultural and Consumer Protection Act of 1973, Pub. L. No. 93-86, 87 Stat. 221; Food and Agriculture Act, Pub. L. No. 95-113, 91 Stat. 913; Agriculture and Food Act of 1981, Pub. L. No. 97-98, 95 Stat. 1213; Food Security Act of 1985, Pub. L. No. 99-198, 99 Stat. 1354; Food Agricultural Conservation and Trade Act of 1990, Pub. L. No. 101-624, 104 Stat. 3359; Federal Agriculture Improvement and Reform Act of 1996, Pub. L. No. 104-127, 110 Stat. 888; Farm Security and Rural Investment Act of 2002 Pub. L. No. 107-171, 116 Stat. 134; Food, Conservation, and Energy Act, Pub. L. No. 110-234, 122 Stat. 923 (2008); Agricultural Act of 2014, Pub. L. No. 113-79, 128 Stat. 649 (citing Laurie Ristino and Gabriela Steier, Losing ground: A clarion call for farm bill reform to ensure a food secure future, 42 *Colum. J. Envtl. L.* 59, 60 [2016]).

[14] Michael Roberts, *Food Law in the United States* at 5 (2016).

of reach of environmental or other statutes that may curb production and, consequently, profitability. Taking a different angle than merely that of production, however, shows that the Jeffersonian ideals of an agrarian nation[15] quickly turned into a USDA-led trend to "get big or get out," where farmers became tools of industrial agriculture,[16] scaling up production and scaling down ownership by joining the bandwagon of standardizing their farms, production, machinery, and, most importantly, their crops by buying seeds from the few BigAg suppliers. Simply put, monocultures of patent-protected GMOs have displaced agroecology (see the second circle on the right in Figure 2.1). Consequently, the variety of crops was reduced to what we now know as staple or commodity crops—wheat, corn, rice, and soy—that took the place of agrobiodiversity.[17] Once again, this development was only made possible under the protective umbrella of private laws, such as the PPPA and USDA regulations furthering the agricultural sector.

Since the passage of the PPPA, the displacement of agrobiodiversity through a handful of commodity crops has focused the market, leading to an expansion in international trade. Although the market shifts and intricacies of international food trade are beyond the scope of this book, it is important to note that the internationalization of the market led to a centralization of the food system by concentrating the power of seed production to the patent holders, a few BigAg moguls that seized control over food production from farm to fork.[18] Farming with tailor-made codependent pesticides and fertilizers, such as Round-Up and Round-Up Ready Corn, and harvesting with sister-company-fueled machinery, processing the crops into widely advertised fast food, and even dumping those on the developing world contributed to the growth of the commodity crop sector.[19] In a type of industrial nepotism, a self-perpetuating reciprocity between the seed industry and the centralization of the food supply fastened BigAg's network (see bottom of Figure 2.1).[20] Conversely,

---

[15] William S. Eubanks II, A rotten system: Subsidizing environmental degradation and poor public health with our nation's tax dollars, 28 *Stan. Envtl. L.J.* 213, 218 (2009).

[16] Id. at 223–225.

[17] Agrobiodiversity is defined as "the result of natural selection processes and the careful selection and inventive developments of farmers, herders and fishers over millennia." It comprises "animals, plants and micro-organisms that are used directly or indirectly for food and agriculture, including crops, livestock, forestry and fisheries," including the diversity of non-harvested species that support production (soil micro-organisms, predators, pollinators), and those in the wider environment that support agro-ecosystems (agricultural, pastoral, forest and aquatic) as well as the diversity of the agro-ecosystems." FAO, Agrobiodiversity (1999), http://www.fao.org/docrep/007/y5609e/y5609e01.htm.

[18] See generally, Mary Angelo, Jason Czarnezki, and Bill Eubanks, *Food, Agriculture, and Environmental Law* (Environmental Law Institute) Environmental Law Institute. Kindle edition.

[19] Id.

[20] Cornucopia Institute, Seed Companies Network.

BigAg has overcome many legal obstacles, ranging from local, national, and international barriers, to become bigger, faster.[21] In the process, the agricultural industry's lobby has exerted political pressures to create market distortions and take a hold of free trade to ensure its continued proliferation. As part of this growth, the food system has become standardized, relying on GMO commodity crops.[22] As an important side effect of the standardization and proliferation of GMOs, their risks have been trivialized (see Chapter 3).

Although policy responses to this rapid expansion of GMO-dependent agricultural growth and its internationalization are ubiquitous, the law has yet to catch up. The augmentation and reduction of regulatory and market barriers to BigAg's growth, as Chapter 3 shows, coincided with industrialization and the consequential urbanization in the nineteenth and twentieth centuries and created pollution, the side effect that links the proliferation of GMOs and the trivialization of its risks to climate change.[23] Figure 2.1 illustrates this circular relationship and links GMOs, urban agriculture, and climate change. Each circle illustrates a point of possible attack, where the law can disrupt the cycle and where a fresh perspective can flip a switch toward improved food integrity. A prerequisite to create any purposeful disruption, however, is the evaluation of whether private law has sufficiently protected food or whether public law control is needed to relieve the tension[24] between GMOs versus agroecology.

Two major Western legal regimes, the United States and the EU, are instructional in exploring the GMOs versus agroecology debate from a legal point of view. Comparing the US and the EU approaches to regulating GMOs reveals the spectrum of issues that the law can control. The reverse is also true: where the laws of each respective regime fail to address GMOs, implicit effects on agroecology and climate change mitigation arise. In a simplified summary, where US private law protects the growth of the GMO-dependent food sector, trivializing the dangers of GMOs on behalf of economic growth, the EU's public laws somewhat prohibit uncontrolled proliferation of GMOs at the expense of food integrity. Thus, the following comparative examination of these seemingly opposite legal approaches and the underlying policy responses to the globalized commodity crop market and climate change challenges of our day and time illuminate the points of attack and relationship that may be ripe for change.

---

[21] William S. Eubanks II, A rotten system: Subsidizing environmental degradation and poor public health with our nation's tax dollars, 28 *Stan. Envtl. L.J.* 213, 218 (2009).

[22] Mary Angelo, Jason Czarnezki, and Bill Eubanks, *Food, Agriculture, and Environmental Law* (Environmental Law Institute) Environmental Law Institute. Kindle edition (2013).

[23] FAO, Agrobiodiversity (1999), http://www.fao.org/docrep/007/y5609e/y5609e01.htm.

[24] Conversation with Prof. Junker (on file with author).

## 2.2   Looking over one's shoulder: The comparativist method

### 2.2.1   What is true? Where are the facts?

Comparative law is interdisciplinary and its methodical application follows a specific pattern,[25] mimicked in this book. Formulating the research question and choice of legal systems are the first steps, followed by a description of the laws of the two countries that are compared.[26] Then, the actual comparison follows, exploring similarities and differences, which are subsequently evaluated.[27] Finally, the analysis based on the comparison may lead to policy recommendations or other conclusions that round off the comparative legal method.[28]

In this book, the research question is how the risks of GMOs for the food system are trivialized through the proliferation of GMOs and how food integrity may be achieved. The two legal systems chosen for comparison are the US-American and the EU regulation of GMOs. At the core of the comparison are the poles-apart US-American protectionist approach and the EU precautionary principles in GMO regulation. Important similarities in challenges for GMO regulation in the United States and the EU are the challenges stemming from urbanization, globalization, and climate change. Finally, the book concludes with lessons that either legal system may draw from the other to find solutions to common problems.

Practicing a comparative legal method, this book juxtaposes selected aspects of US-American and EU food law, regulation, and policy, with the goal to highlight how one system can learn—or borrow solutions to shared problems—from the other. This functional comparison, that is, one that asks *what* is being compared and where "the core element … [is] a socio-economic problem … [as] the starting point of a comparative analysis,"[29] is essentially the perspective shift previously described in the analogy of the chess player challenging himself in a match. Here, "the shared purpose of these rules is the common denominator (*tertium comparationis*) which allows comparability of these legal systems" and is the problem of GMO regulation in the United States and the EU.[30] Although scholars propose that "one can start with the way different legal systems deal with a particular problem and then compare these approaches in

---

[25] Mathias Siems, *Comparative Law (Law in Context)* (p. 13). Cambridge University Press. Kindle edition.
[26] Id.
[27] Id.
[28] Id.
[29] Mathias Siems, *Comparative Law (Law in Context)* (p. 26). Cambridge University Press. Kindle edition.
[30] Id.

terms of economic efficiency,"[31] as is often done, this book compares the US-American and EU approaches of GMO regulation in terms of their potential to achieve food integrity.

The assumption made at this point is that law, in either system, has a role outside of its connection to economics, rather than only existing to serve economics.[32] Elemental to this assumption that economics need not be the sole driving force of law and, inherently, of policy that leads to law is that moral virtues and nonmonetary values are fundamental to food integrity and that they matter. This assumption, in turn, follows the big-picture path of the United Nations' analysis in the numerous publications on the Sustainable Development Goals, the Declaration of Nyeleni, and the Right to Food.[33]

Resuming the question of achieving food integrity, the "'praesumptio similitudinis,' a presumption that the practical results are similar," as Zweigert and Kötz formulated in their seminal work on comparative law,[34] confirms that the facts and questions about GMO regulation are the lowest common denominator in the examination of the US and the EU approaches. The chess pieces are the common facts underlying the analysis in this book, such as measurable climate change impacts, urbanization data, and scientific trends. Arguably, the pieces do not change; only their position and use shift strategically. In other words, the relationships of the functional parts in the food system, as defined by Professors Lawrence and Neff in Chapter 1, change when the parts are moved—just like the metaphorical pieces on a chess board. Facts, in comparative law, present "[t]he real difficulty ... to distill the raw information into a measure that achieves a close fit between the facts and the concept (validity) in a reproducible, consistent manner (reliability)."[35] This is why GMOs present an excellent angle for comparison: the underlying facts are hotly disputed and seem unreliable at first sight, but they, in fact, share some common legal enablers.[36]

---

[31] Id.

[32] Correspondence with Prof. Junker (Apr. 1, 2017) (on file with author).

[33] Mathias Siems, *Comparative Law (Law in Context)* (p. 26). Cambridge University Press. Kindle edition.

[34] Mathias Siems, *Comparative Law (Law in Context)* (p. 30). Cambridge University Press. Kindle edition (citing Zweigert and Kötz, *supra* note xx, at 40).

[35] Holger Spamann, Empirical comparative law, Discussion Paper No. 815 Harvard Law School, published in *Annual Review of Law and Social Science* Vol. 11 (2015), http://www.law.harvard.edu/programs/olin_center/papers/pdf/Spamann_815.pdf (last accessed May 29, 2017).

[36] Angelika Hilbeck et al., No scientific consensus on GMO safety, *Environmental Sciences Europe* (2015) 27: 4, https://goo.gl/W4evoS (last accessed May 29, 2017); *see also* Debating Europe, Arguments for and against GMOs (undated), http://www.debatingeurope.eu/focus/arguments-gmos/#.WSw-8VKZNE4 (last accessed May 29, 2017).

Similarly, comparison must follow strategy and choose which facts are justifiably comparable. Truth and facts, as their plain meaning implies, are observable, actually existing, and true. As Professor Dr. Kirk Junker from the University of Cologne explains, facts have been stable because they come from science.[37] However, he warns that the understanding of facts is changing, and, therefore, one has to show how facts and knowledge are established.[38] Simply put, to explain what should be taken as factual requires definition. For the 1993 Presidential Address of the North American Conference on British Studies, Barbara Shapiro explained that the concept of working with facts "is an adaptation or borrowing from another discipline-jurisprudence, and that … fact-finding in law were carried over into the experimental science…."[39] In legal history, matters of fact and matters of law were distinguished,[40] and they boil down to an inquiry for fact-finders, such as jurors in the US legal system, and the judges, who apply the law to the facts. Therefore, this book relies on definitions (see Glossary) promulgated by the United Nations or accepted widely in the legal community, so as to ensure that the facts in the US-EU comparison are as close as possible.

How facts and law come together, however, depends on context, that is, the relationship between the facts, as comparative law scholar, Mathias Siems explains.[41] He notes that "comparative law broadens the understanding of how legal rules work in context."[42] Such context may, for instance, be a globalizing trend,[43] such as the proliferation of GMOs in food systems. Conversely, culture often provides context. In food law specifically, facts are often disputed and cultural discrepancies divide opinions. The crux of the GMO debate, for instance, villainizes pesticides and synthetic fertilizers upon which industrial-scale monocultures depend. Stuart Newman, Professor of Cell Biology and Anatomy at New York Medical College, explains that the idea of "improving foods and other crop plants by introducing foreign genes was among the first applications proposed" as soon as genetic modification became possible through gene transfer.[44] Once people recognized the risk of creating a cancerous organism with the mutation of just one gene, "concerns were raised about the capability of the transgenic methods to dramatically change the biochemistry

[37] Kirk Junker, Doktorantenseminar January 2017 (notes on file with author).
[38] Id.
[39] Barbara Shapiro, The concept of "fact": Legal origins and cultural diffusion, *Presidential Address of the North American Conference on British Studies* (1993).
[40] Id.
[41] Mathias Siems, *Comparative Law (Law in Context)* (p. 3). Cambridge University Press. Kindle edition.
[42] Id.
[43] Id.
[44] Stuart Newman, The state of the science, *in The GMO Deception* (Sheldon Krismsky and Jeremy Gruber, Eds.). Kindle location 653.

or ecological stability of plants."[45] Fear of allergenicity, toxicity, or eco-system disruptions through "superweeds," began to dominate the GMO debate.[46] In the United States, consumer watchdog organizations vigor-ously defend this anti-GMO angle of the debate. In contrast, commercials, lobbyists, and "revolving door" lobbyists shape the pro-GMO discourse with craft. In the EU, the cultural context is different, where consumers and the media take a somewhat anti-GMO approach.[47] What this cultural discrepancy shows is that the possibility to transfer genes was the fact, but the effects gave rise to perspective shifts. On the one hand, BigAg saw opportunities for immense growth,[48] while environmentalists and consumer watchdog organizations realized the risks that GMOs have for the food system in the United States.[49] On the other hand, the EU's public and media took a wait-and-see approach, observing the development of GMOs prior to accepting them.[50] How the genetic manipulation of crops was to be positioned upon the metaphoric chessboard thus became a question of perspective.

In the GMO debate in comparative food law, two perspectives are seemingly poles apart: the EU and the United States. Although Chapters 3 and 4 discuss the European precautionary approach, as compared to the US biotech principle, one idea must be highlighted here: The existence of GMOs and the factual effects create problems for both jurisdictions, which constitute an important commonality and illustrate that the law has yet to catch up, as previously noted in Chapter 1. Perhaps examin-ing each other's problems can lead to borrowing solutions that may be adapted to one another's needs.

## 2.2.2    *Legal opinions and perspectives: Why borrow?*

Why borrow?[51] Professor Alan Watson from the University of Georgia School of Law explores this question and gives various reasons. First, he

---

[45] Id.

[46] Id.

[47] These cultural distinctions are based on the author's decade-long experience living in both the United States and the European Union and her firsthand observations.

[48] Monsanto, What is a GMO?, http://discover.monsanto.com/monsanto-gmo-foods (last accessed Mar. 9, 2017).

[49] Center for Food Safety, Are GMOs safe?: No Consensus in the Science, Scientists Say in Peer-Reviewed Statement (Feb. 19, 2015), http://www.centerforfoodsafety.org/press-releases/3766/are-gmos-safe-no-consensus-in-the-science-scientists-say-in-peer-reviewed-statement.

[50] See *supra*.

[51] Alan Watson, Legal culture v. legal tradition, in *Epistemology and Methodology of Comparative Law* at 3 (Mark Van Hoecke, Ed.) Hart Publishing (2001); Vlad Perju, Constitutional trans-plants, borrowing and migrations, in *Oxford Handbook of Comparative Constitutional Law* (Michel Rosenfeld and Andras Sajo, Eds.) Oxford University Press (2012).

writes, "it is easier to borrow than to create rules and institutions from new."[52] Second, and more importantly, borrowing satisfies the need for authority,[53] that is, the power to adjudicate and settle disputes. In the US common law system, authority often comes from precedent, where courts apply common law and statutes to facts to settle disputes. By comparison, in the civil codes of most EU member states, authority comes from the statutory language itself and its application—with little regard for precedent. Notably, however, the federalism of the United States and its supreme court precedent are rarely being borrowed.[54] Nonetheless, the US protectionist set of laws pertaining to GMO regulation may be singled out to help the interdisciplinary problem of GMOs versus agroecology. The *how* and *how not* to regulate GMOs in the United States, evaluated in terms of the potential to achieve food integrity, is instructional beyond the mere letter of the law. Simply put, this book examines the effects of these laws rather than their statutory language.

In order to examine the effects of the law, the general structure of the US adversarial system is notable in understanding the context that is vital for comparativists. Under the adversarial design, seeking a "winner" who prevails by arguing more strongly is a large part of the US fact-finding and judicial system. By comparison, the continental European counterpart is inquisitive, seeking the truth in what is being argued. That truth, ideologically, prevails in a lawsuit. In other words, common law[55] is the "body of law derived from judicial decisions, rather than from statutes or constitutions," while the European *"civil law* (commonly referred to as *jus civile*) denotes the whole body of Roman law, from whatever source derived."[56] Thus, justifications for legal opinions are grounded in the authority that judges, lawyers, and legal scholars borrow from other sources of authority, which are, of course, neither limited to precedent nor codes. In fact, some systems lend themselves to be quite suitable donors[57] in the cases of regulation and legislation about food, and, particularly, GMOs. Although Ireland and the United Kingdom are both European and common law systems, this book focuses on the United States and the EU as each other's counterparts in this comparison of the GMO versus agroecology debate. Extrapolating other nuances in the relationship between the legal system and the perception of the GMO debate in other countries is beyond the scope of this book.

[52] Watson, *supra* note 1 at 3.

[53] Alan Watson, Legal culture v. legal tradition, in *Epistemology and Methodology of Comparative Law* at 3 (Mark Van Hoecke, Ed.). Hart Publishing (2001).

[54] Adam Liptak, U.S. court is now guiding fewer nations, *The New York Times* (Sept. 17, 2008), http://www.nytimes.com/2008/09/18/us/18legal.html?hp (last accessed May 30, 2017).

[55] Common Law, *Black's Law Dictionary* (10th ed.; 2014).

[56] Civil Law, *Black's Law Dictionary* (10th ed.; 2014).

[57] Watson, *supra* note 1.

The shared problems that transcend national borders and jurisdictions, which are described in this book, may also share solutions. For instance, the EU can learn from the United States, both from its achievements as well as its failures in promoting GMOs. Certainly, the reverse is true, where the United States can learn from the EU's precautionary approach. As Professor Watson framed it, when one solution "is not available in their own system, they seek it elsewhere...."[58] It is through this method of borrowing lessons and solutions that this book proposes to tackle the legal problems that the proliferation of GMOs has brought about for the US and the EU food systems.

A prerequisite of borrowing is assuming the same root cultures, which the EU and the United States have to some extent from a historical perspective and if compared at large to other continents. The assumption made in this book is that the culture of consumer protection in the United States resembles that of the EU and that principles of democracy, capitalism, and free trade suffice to allow a borrowing, which remains, of course, subject to adaptation.[59] As such, the author recognizes what Günther Frankenberg coined as the basic contention against the work of comparativists, "that ... is necessarily laden with concepts, values and visions derived from their local legal culture and experience."[60] This author, having lived and studied in both the United States and the EU, and being sensitized to cultural differences and similarities, follows Frankenberg's "hope to circumvent the problem of perspective by positing [oneself] as [a] natural observer," instead of coping with ethnocentrism.[61] The perspective shifts in this book are intended to help maintain objectivity and derive benefits for the reader from the comparative method because "we can benefit from the perspectives of others who react to, and analyze its functionality."[62] On the same note, Professor Dr. Junker once analogized legal comparison to learning a language. He writes:

> We learn our native language by usage, and learn
> that usage in the context of its native spoken culture.
> We learn additional languages through the methodi-
> cal mechanics of grammar and vocabulary, outside
> of the native cultural context. Therefore, we should

---

[58] Id.

[59] See, generally, Jessica Watts, The transatlantic trade and investment partnership: An overly "ambitious" attempt to harmonize divergent philosophies on acceptable risks in food production without directly addressing areas of disagreement, 41 *N.C.J. Int'l L.* 83 (2015); David A. Wirth, The EU's new impact on U.S. environmental regulation, *Fletcher F. World Aff.* at 91. Summer 2007.

[60] Frankenberg, *supra* note xx, at 442.

[61] Id.

[62] Kirk W. Junker, A focus on comparison in comparative law, 52 *Duq. L. Rev.* 69, 78 (2014).

be wary of the mechanical artificialities, out of context, when learning additional legal cultures and making comparisons. Even the native who does not study the law as a legal specialist has learned the values, norms and procedures in that law, at some level, from everyday experience in his or her culture. Some of the things a native of any culture learns about his legal system might even be regarded by legal professionals as being wrong, but if it is generally learned in that culture, it is a force that forms general cultural expectations among the natives and cannot be dismissed. Equally important is the fact that one should not study a foreign society's law, out of cultural context, as though it is just another substantive law course or practice area in one's own system.[63]

Keeping Professor Junker's aforementioned analogy of learning language in mind, this author embarks upon exploring the issues of the following chapters with the goal to show readers how these issues factor into international comparative food law. Acknowledging all the pitfalls that comparison has, this author notes that she calls both the United States and continental Europe her home, hoping that she is sufficiently culturally sensitized to avoid the traps of bad comparison. From experience, she highlights that issues that matter in one culture may not matter as much in the other, and vice versa.

## 2.3   Comparison as a form of crossborder knowledge acquisition[64]

> "It is difficult to understand the universe if you only study one planet."[65]
>
> **Miyamoto Musashi,**
> *A Book of Five Rings*

Acknowledging the multifaceted and varied field of comparative law, this book is focused on the study of legal transplants, the aforementioned borrowing. Over three decades ago, Günther Frankenberg's famous article published in the *Harvard Journal of International Law* in 1985 already acknowledged the perspective as a "central and determinative element

---

[63] Id. at 77.
[64] Jaakko Husa, *A New Introduction to Comparative Law*, at 58. Bloomsbury (2015).
[65] Cited in Jaakko Husa, *A New Introduction to Comparative Law* (2015).

in the discourse of comparative law,"[66] highlighting the importance of perspective as described through the turning chessboard analogy in Chapter 1.[67]

How one chooses to turn the board and flip the switches, however, is a matter for comparative method. Professor Vivian Grosswald Curran from the University of Pittsburgh School of Law highlights the complications for comparativists in choosing methods, such as the universalist tradition, translation, or cultural immersion.[68] To some scholars,

> the borrowing of ideas between legal cultures and/ or systems [is] the most fruitful way of exploring the relationship between law and society, and the underlying perceptions of law; as well as the magnifying glass through which one best observes how state law lives side by side with other (supranational and domestic) sources of law and, thereby, how relative the notion of state power … can be.[69]

Thus, in an effort to cope with the multiplicity of law and to make the analysis more digestible, this book narrows its scope both from an international view not just to the EU and the United States, but also to the specific field of food law as it relates to food safety and environmental protection. Moreover, to zoom back out from the *tertium comparationis*, this book uses the approach referred to as *da lege lata/da lege lata*,[70] that is, from broad law to broad law. This forced simplification allows the borrowing of ideas between the legal cultures of the EU and the United States, focusing on federal US and EU-wide law and policy, thereby contextualizing the legal regimes with societal, economic, and political underpinnings, while allowing some cultural understanding to be factored into the analysis.

---

[66] Günther Frankenberg, Critical comparisons: Re-thinking comparative law, 26 *Harvard Journal of Comparative Law* 2 (1985), http://iglp.law.harvard.edu/wp-content/uploads/2014/10/Frankenberg-Critical-Comparassons-excerpt.pdf.

[67] Here, the legal transplants also transcend disciplines, thereby uniting agriculture, urban planning, food science, policy, and law into one line of argument.

[68] See generally, Vivian Grosswald Curran, *Comparative Law: An Introduction.* Carolina Academic Press (2002).

[69] Diapositives versus movies: The inner dynamics of the law and its comparative account, in *The Cambridge Companion to Comparative Law* at 3 (Mauro Bussani and Ugo Mattei, Eds.). Cambridge University Press (2012).

[70] Juha Karhu, How to make comparable things: Legal engineering and the service of comparative law, in *Epistemology and Methodology of Comparative Law* at 80–81 (Mark Van Hoecke, Ed.). Hart Publishing (2001).

## 2.4    Law as a function of culture:[71] Borrowing norms and assuming similarity

How can one culture just borrow norms from another as though the underlying cultural norms are the same and will support the same rules of law?[72] Professor Junker teaches that "analyses can only be conducted if the questions used are capable of yielding similarities,"[73] which is a principle that Zweigert and Kötz coined as "*praesumptio similitudinis* (presumption of similarity)."[74] As such, this book follows Zweigert and Kötz's "replicable method of functionalism which one might employ as a social scientific tool to achieve respectable comparisons"[75] while attempting to circumvent the thought fallacies that ethnocentrism, as explained by Günther Frankenberg, may create.

One form of presuming similarity rests in legal harmonization, which, in turn, requires unification of authority. Notably, some systems are more authoritative than others.[76] Conceding that "neither authority in the social sense nor authority in the legal sense are easily transferrable from one culture to another culture,"[77] the shared historic roots of the foremost Western legal cultures, the European and the US-American ones, are obvious reinforcement for borrowing from each other. For instance, in light of the negotiations over the Transatlantic Trade and Investment Partnership (TTIP),[78] a trade and investment deal between the EU and the United States, this authority will be further supported. The TTIP will supposedly set "new rules to make it easier and fairer to export, import and invest overseas," open up the US market to European export, and give the EU and the United States "influence [over] world trade rules" and project their values globally.[79] Thus, commercial harmonization following the US model is supposed to be transplanted into the EU—a fantasy overlooking the rest of the multiplicity of European culture that generally resists the American capitalist and biotechnology approaches.[80] Critics warn that the European food system will suffer in quality, safety, and sustainability

---

[71] This phrase is borrowed from Kirk Junker (notes on file with author).

[72] Kirk Junker, *supra* note lxi.

[73] Kirk W. Junker, A focus on comparison in comparative law, 52 *Duq. L. Rev.* 69, 80 (2014).

[74] Id. (*citing* Konrad Zweigert and Hein Kötz, *Introduction to Comparative Law* 6 [Tony Weir trans., 3rd ed. 1998]).

[75] Kirk W. Junker, A focus on comparison in comparative law, 52 *Duq. L. Rev.* 69, 82 (2014).

[76] Watson, *supra* note l.

[77] Kirk Junker (notes on file with author).

[78] Europa, Transatlantic Trade and Investment Partnership (TTIP), http://ec.europa.eu/trade/policy/in-focus/ttip/about-ttip/ (last accessed Jul. 1, 2016).

[79] Id.

[80] Debbie Barker, BigAg's wish list fulfilled in TPP, Center for Food Safety (Nov. 14, 2015), http://www.centerforfoodsafety.org/blog/4123/big-ags-wish-list-fulfilled-in-tpp (last accessed Mar. 9, 2017).

if the United States imposes their values, provoking clashes between the precautionary and biotechnology principles, discussed in Chapters 3 and 4.

The TTIP illustrates, however, the point at which legal culture and legal tradition intersect, showing that borrowing does not always work. "Borrowing is only part, though perhaps the most obvious, of the conjunction of legal culture and legal tradition,"[81] explains Professor Watson. The search for justification with one's own legal system, he continues, is inherently cultural and "inevitably backward looking."[82] Thus, authority, as essential as it is for law, creates the legal tradition, which consists of the actual conjunction of legal borrowing and the need for authority.[83] This also explains why the TTIP was well received in the United States under former president Obama's administration, but less so under Trump, where food and environmental law are governed by a "safe until proven unsafe" approach. However, in the EU, the TTIP is most viciously contested because the "better safe than sorry" principle reigns. Ultimately Professor Watson's analysis applies directly to the TTIP negotiations and the GMO regulatory framework of this book: "The startling and upsetting conclusion is that a system of private law must be understood primarily in terms of its own legal history, not societal, political and economic history in general."[84] This is where food law in the United States and the EU runs afoul of each other, and where the remainder of this book picks up.

Conceding that even the comparative method used herein has its limits, it does not go unnoticed that every legal comparison needs a yardstick or a common denominator, the previously mentioned *tertium comparationis*.[85] The difficulty, however, lies in the comparison of only selected qualities from different perspectives.[86] As with other disciplines, the comparative approach to food law is also built on the fundamental commensurability,[87] a shared set of problems addressed through the law of the compared entities, here, the EU and US frameworks of GMO regulation. This commensurability, however, does not presume similarity.[88] The differences enable comparison through the *tertium comparationis*, which refers to a common quality that two compared entities share.[89] In this book, the common denominator is a combination of (1) shared features, the EU and US regulation of GMOs generally, legal frameworks that

[81] Watson, *supra* note l.
[82] Id.
[83] Id.
[84] Id. (internal citations omitted).
[85] Jaakko Husa, *A New Introduction to Comparative Law* at 148. Bloomsbury (2015).
[86] Id.
[87] Id.
[88] Id.
[89] Watson, *supra* note l; see also Husa, *supra* note lxiii.

are only just emerging on the international stage; and (2) shared functions, the results of each federal framework's regulation on food safety and environmental integrity. Comparison being an absolute necessity for this book, the comparative framework is set by the author.[90] As such, the methodology, or yardstick, chosen here is, as with many comparative legal approaches, "a question of the epistemic point of view taken to the objects chosen for study; from the framework of approach to the theme."[91] More specifically, the yardstick here is the achievability of food integrity in either system. By way of analogy, apples and oranges cannot be compared, but their nutritional profiles, levels of acidity, colors, and textures can be compared. The same is true for EU and US law, which seems to be as vastly different as apples and oranges.[92] However, the scale chosen for this comparison is that of food safety and environmental integrity. Thus, the two subjects of comparison are US and EU food laws. Anything that jeopardizes either food safety or environmental integrity in these frameworks is the *tertium comparationis* of this book, limited by the scope chosen by the author. Logically then, the shared US and EU features are the tasks of regulating food safety and environmental integrity through the law in either system. Correspondingly, the shared functions lie in the protection of food safety and environmental integrity.

As the quote introducing this subsection states, the universe cannot be understood based on just one planet. The same can be translated to the realm of food law. The modesty of this book lies in the concession that the universe of food law, generally, is made of so many metaphorical planets that only some can be chosen for the comparison here. Nonetheless, this inward-oriented approach, zooming in from EU and US comparative law, to the comparison of only food and environmental law, and then again focusing on a handful of issues, can be quite instructive. Such selective comparison, in fact, helps to understand the bigger picture, the global scale that GMO trade has, the implications of the larger principles governing EU and US food law and policy, and the potential dangers looming in trade partnerships. Essentially, comparison is the turning of the chessboard to shift perspective in the quest for how solutions can be borrowed for shared or overlapping problems.

## 2.5   Summary

This chapter shows how the proliferation of GMOs creates food insecurity by denying people's access to food through food system centralization, and how industrial agricultural policy to-date largely trivializes the

---

[90] Husa, *supra* note lxiii.
[91] Id.
[92] Id.

dangers of GMO monocultures to crop diversity and biodiversity, thereby, weakening food systems. All of these negative tendencies are made possible through private law and the protection it awards to those controlling the food system. The legal analysis identifies points of attack to improve upon the current food system using public law as a tool, that is, to pierce the shroud of private legal protection that creates the GMO-centered enabling system. On its flip side, the legal analysis also highlights where the effectiveness of legal tools end and opportunities for solutions to problems of food insecurity, trade imbalances, and other challenges begin. This chapter uses a perspective switchboard, where a shift in viewpoint uses a comparative legal examination of the facts linking GMOs, agroecology, and urban agriculture.

# chapter three

# Food dependence: GMOs fail to feed the world

Food dependence is the opposite to food security and to food sovereignty, and it describes people's reliance on and their confidence and trust in food producers, contingent upon the farming practices of the industry and all of its components. Simply put, food dependence relinquishes control over how to obtain one's food in the broad sense. Examples of food dependence are the inability to save seeds for future planting and the inability to find affordable produce in food deserts where only fast and processed food is available. The commonality of the various forms of food dependence are distortions of facts, which Section 3.1 details. Zooming in on one specific example of food dependence is the outsourcing of agriculture in metropolitan areas. Thus, Section 3.2 explains how one may feed a city in light of the difficulties that the proliferation through trivialization of genetically modified organisms (GMOs) has created. Then, Section 3.3 explores a solution to insource food, by using the vacant rooftop space in cities to plant fresh fruits and vegetables, thereby transforming urban food deserts into lush landscapes that nurture the communities in cities.

## 3.1  Debunking the myths of GMO farming

### 3.1.1  Lowest common denominator and consumer deception

This section targets consumers and those protecting their right to know where their food comes from—information that GMO producers often shroud in secrecy.[1] From a legal perspective, this section describes the effects of the US-American protectionist approach based in private law, which is later contrasted with the EU's public law approach to GMO regulation. At the onset, however, the cultural context will be examined.

Mainstream US-American consumers generally use food culturally rather than nutritionally. For instance, powerful advertising seduces consumers to eat out of boredom. The fact that lack of fitness reduces the ways in which individuals can get pleasure also leads them to eat, overeat, and

---

[1] See generally, Sheldon Krismsky and Jeremy Gruber, *The GMO Deception: What You Need to Know About the Food, Corporations, and Government Agencies Putting Our Families and Our Environment at Risk*. Skyhorse Publishing. Kindle edition (2014).

yet suffer from low energy. Additionally, social sharing focused on food, such as "let's get a coffee," or "let's get a beer," or "let's get lunch" being used to mean "let's meet," further illustrates how the US-American public uses food.[2] Moreover, the average American eater, who "consum[es] 24 lbs of artificial sweeteners, 29 lbs of French fries, and over 600 lbs of dairy per year"[3] is malnourished and misinformed. Virtually no American meets the government's dietary goals, which are already a watered-down, lowest common denominator of what might constitute a healthy and balanced diet, exceeding the recommended amounts of sodium (salt), saturated fat, and sugars.[4] Moreover, these average consumers are influenced by food advertisements in popular culture and think that what tastes "good" *is* good. They trust the government to regulate restaurants and supermarkets to only allow safe food. Finally, these average consumers are ignorant as to where their food comes from and how it is made—an observation rooted in the protectionism of the food industry and the private rights awarded to it under US law.

Other than their European counterparts, average US consumers are decreasingly aware of the effects that their food has. They may, for instance, feel their minds' drifts to possible snacks between meals and sometimes feel controlled by cravings.[5] It is because their diet fails to provide adequate nutrition. This means that they get full, but not nurtured. The food they eat is simply not satisfying because it does not give their bodies the fuel they need to be active and healthy.[6] Unfortunately, they do not live in a society where they can afford to relinquish control over our food choices to others. The US government, for instance, has allowed lobbyists working for fast-food companies to have Congress classify pizza, chips, and French fries as vegetables.[7] Doctors and nutritionists generally only learn the dietary recommendations offered by the government, which is biased toward protecting industrial agriculture and the processed fast foods that come from it.[8] Magazines and television programs are paid to

[2] Correspondence with Professor Junker (on file with author).

[3] Mike Barrett, Average American Diet—Infographic, Natural Society (May 17, 2012), http://naturalsociety.com/average-american-diet-infographic/.

[4] Dietary Guidelines 2015–2020, Current Eating Patterns in the United States, Health.gov, https://health.gov/dietaryguidelines/2015/guidelines/chapter-2/current-eating-patterns-in-the-united-states/.

[5] Sara Miller, The Science of Hunger: How to Control It and Fight Cravings, *Live Science* (Apr. 1, 2016), http://www.livescience.com/54248-controlling-your-hunger.html.

[6] FAO, FAO's role in nutrition (undated), http://www.fao.org/nutrition/en/.

[7] Earl Blumenauer, Congress Says Pizza and French Fries Are Vegetables!, *Huffington Post* (Nov. 16, 2016), http://www.huffingtonpost.com/rep-earl-blumenauer/congress-says-pizza-and-f_b_1098207.html.

[8] Marion Nestle, No nutrition in medical education? An old story that might be changing, *Food Politics* 8 Apr. 1, 2014, http://www.foodpolitics.com/2014/04/no-nutrition-in-medical-education-an-old-story-that-might-be-changing/.

make food look good that is not.[9] There is a lot more to this, and it means that the food industry is tricking consumers' minds into thinking they are eating to satiate their appetites by manipulating their senses and stripping them of control over their decision about what to put into their bodies.[10] All of this seemingly unfair set of practices, albeit common, is not illegal. On the contrary! US private law enables these food system distortions.

### 3.1.2   Food choices have a ripple effect on the environment

Another consideration that many consumers overlook is that their food choices have a ripple effect on the environment and the economy. Considering the big picture, farming (i.e., food production) uses common resources, such as fresh water, soil, and other resources. Depleting and polluting our environment through intensive GMO monoculture farming affects the common resources,[11] which, if harmed, reduce the supply for further food production. As such, permitting unsustainable farming to harm not only public health but also the environment requires legal protection. As Garret Hardin described in his seminal work about the tragedy of the commons, there is no "wild hope that improved food production technology will allow an indefinite increase in population" because "a finite world can support only a finite population."[12] Thus, food production that fails to mitigate climate change and work in harmony with the environment, that is, agroecology, points toward the tragedy of the commons, where resources are depleted and, in the extreme case, lead to starvation. Consequently, environmental conservation and climate change mitigation are integral parts of food production, striving to stop food insecurity and dependence, thereby supporting the premise that agroecology is a path toward food integrity.

On the same note, as Marion Nestle, bestselling author of *Food Politics* and professor of nutrition, food studies, and public health at New York University illustrates in her latest book, *Eat, Drink, Vote*, eating is a political act that could give consumers control back over their diet and the environmental footprint of their meals, thereby reducing the impact on climate change.[13] Food choices ultimately relate to the air we breathe, the water we drink, and the education our children obtain. Even the United Nations (UN) warn that,

---

[9] Michael Moss, The Extraordinary Science of Addictive Junk Food, *Time Magazine* (Feb. 20, 2013), http://www.nytimes.com/2013/02/24/magazine/the-extraordinary-science-of-junk-food.html.

[10] See generally, Sheldon Krismsky and Jeremy Gruber, *The GMO Deception: What You Need to Know About the Food, Corporations, and Government Agencies Putting Our Families and Our Environment at Risk*. Skyhorse Publishing. Kindle edition (2014).

[11] Garrett Hardin, The tragedy of the commons, *Science* 162 (3859), pp. 1243–1248 (1968), http://science.sciencemag.org/content/162/3859/1243.full (last accessed May 30, 2017).

[12] Myers, http://faculty.wwu.edu/gmyers/esssa/Hardin.html.

[13] See generally, Marion Nestle, *Eat, Drink, Vote*; NRDC, Eat Green: Our everyday food choices affect global warming and the environment, NRDC Fact Sheet (Feb. 2010), https://www.nrdc.org/sites/default/files/eatgreenfs_feb2010.pdf.

climate change, environmental sustainability and
rapid technological shifts, are transforming the food
system and raising questions about how to feed a
growing world population in sustainable ways. At
the same time, uneven economic growth, social
and economic transformations and other factors
are shaping food systems and diets. As a result, the
prevalence of overweight, obesity and related non-
communicable diseases are increasing while under-
nutrition and micronutrient deficiencies persist.[14]

This is why consumers should ask hard questions about their diet
and inquire about the quality and origins of their food. Section 3.2 reveals
more about seizing control over one's diet, health, and consumer choices.

## 3.2   How do you feed a city? Insourcing food security

*Through the years, while my buildings have fallen apart,*
*I've worried about it with all of my heart…*
*UNLESS someone like you cares a whole awful lot,*
*nothing is going to get better. It's not…*
*Plant a new [tree]. Treat it with care.*
*Give it clean water. And feed it fresh air.*
*Grow a forest. Protect it from axes that hack.*
*Then the Lorax and all of his friends*
*may come back.*

**Dr. Seuss,**
*THE LORAX (1971)*

Global population growth reached a tipping point in mid-2009, when
the world population became more urban than rural for the first time
in history.[15] The UN predicts that 60% of the global population will live
in cities by 2050.[16] Logical challenges from this urbanization are institutional capacity and making large cities more livable[17] because cities
have a parasitic ecological footprint. They "consume 75 [percent] of the
world's resources, while covering only 2 [percent] of the world's surface,
and emit 80 [percent] of greenhouse gases."[18] In large cities of over five

---

[14] FAO, *supra* note 4.

[15] Johannes Wiskerke, Urban food systems, in *Cities and Agriculture: Developing Resilient Urban Food Systems* (Henk de Zeeuw and Pay Drechsel, Eds.). Routledge (2015); VitalSource Bookshelf Online.

[16] Id.

[17] Id.

[18] Ash et al., Reimagining cities, 319 *Science* 5864, 739 (Feb. 2008), http://www.sciencemag. org/content/319/5864/739.full.

million inhabitants, "energy demand for mobility, for cooling and heating of houses and offices, for all sorts of equipment for domestic use, and for long-distance transport, processing, packaging, cooling and storage of food" is growing.[19] As such, urbanization is a major culprit in driving a quantity- rather than a quality-oriented agricultural sector toward unsustainable GMO monocultures, which, in turn, are heavily dependent on fossil fuel, deplete natural resources, and contribute to pollution and climate change. All of this is, again, possible due to the power imbalance between private and public law in the food sector. Nonetheless, city greening and urban agriculture are giving renewed hope to reduce the urban environmental harm.[20]

One type of city greening combined with urban agriculture that unites both climate change mitigation and food integrity achievement are urban rooftop farms (URFs). With the proliferation of such URFs, the question arises about who may regulate them and how some balance between private and public law may be restored. To date, urban agriculture is locally regulated by a complex food safety regime. Thus, this chapter proposes that with adequate food security regulation measures, URFs could contribute to city greening and climate change mitigation while providing fresh, locally sourced produce for growing urban populations. First, Section 3.2.1 explores how cities and urban food security evolved and how they factor into ongoing climate change mitigation efforts. Then, Section 3.2.2 explains the benefits of combining urban agriculture and city greening into URFs as part of the trending food movements toward food integrity. Finally, Section 3.2.3 analyzes three regulatory approaches to URFs and illustrates them with examples that unite public and private law efforts.

## 3.2.1  Urbanization of food and agriculture

The separation of agriculture and urbanization in large cities brought about food insecurity as megapolitan cities lost their agricultural independence.[21] With an absurdly insatiable Western appetite and through urbanization, humans have ruined fertile natural landscapes and replaced them with barren, artificial ones.[22] Agriculture was, thereby, outsourced to rural areas. Ancient Rome, for instance, fed nearly one million citizens by importing food via ocean access,[23] expanding the empire to Carthage and Egypt in a "long, drawn-out militarized shopping spree," and even importing oysters from London, as bestselling author Carolyn Steel points

---

[19] Id.
[20] Id.
[21] Steel, *supra* note 1.
[22] Id.
[23] Id.

out.[24] The Roman Empire illustrates how the ability to access food across vast distances accelerated the centralization of cities, effectively exploiting rural frontiers.[25]

Additionally, as demands rose to feed growing populations, food was transported, traded, and imported, thereby gradually increasing food miles.[26] In seventeenth-century London, for instance, meat was imported from Scotland and Wales and brought to the inner-city Smith Field meat market. A map from that time shows how food could be traced to the streets where it was processed and sold, such as on Poultry Hill, Hogg Lane, and Fish Street (see Figure 3.1). As with London, city life in Europe and the United States was shaped by food.[27]

Over time, the presence of food has been marginalized, isolating people from the sources of the food they consume. In the eighteenth-century United States, farm animals were some of the first railway passengers, "being slaughtered out of sight and mind, somewhere in the countryside."[28] Thus, cities were no longer geographically constrained by the newly outsourced food production, and they expanded further.[29] During this urbanization, food became directly dependent on transportation, emancipating cities from nature,[30] but sacrificing transparency for economic growth and booming populations.

Ironically, the modern food systems that evolved, in part through urbanization, complicated matters that were supposed to become easier.[31] Today, city dwellers have become so distanced from nature that they can no longer determine where their food comes from or whether it is safe—they need to be told by colossal and confusing regulatory agencies.[32] By sending agriculture into rural exile, urbanites lost touch with their food, delegating control over food security to the government.[33] Moreover, as Steel notes, these emerging food systems forced people into co-dependency with unsustainable agriculture that large agribusinesses argue only they can deliver.[34] Therefore, in the twenty-first century, recentralization

---

[24] Id.

[25] Id.

[26] A food mile is "a measure of the distance traveled by foods between the place where they are produced and the place where they are eaten. Long distances are considered bad because the quality of the food is worse and energy is wasted in transporting it." *Macmillan Dictionary*—food mile, http://www.macmillandictionary.com/us/dictionary/american/food-mile (last accessed May 30, 2017).

[27] Steel, *supra* note 1.

[28] Id.

[29] FAO, The Multiple Dimensions of Food Security, http://www.fao.org/publications/sofi/2013/en/.

[30] Steel, *supra* note 1.

[31] Id.

[32] Lisa Heinzerling.

[33] FAO, *supra* note 15.

[34] Steel, *supra* note 1.

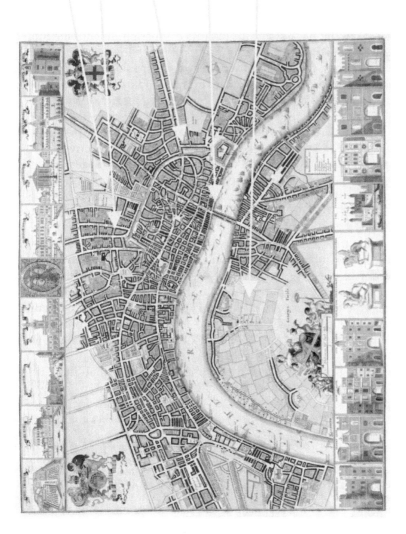

**Selected street names**

Hogg lane

White Crops and Red Crops streets

Cheap Side Poultry and Corn hill, Bread street

Fish Street hill, Lime Street, Fish Freeze hill, and Little Catching

*Several agricultural fields south of London*

*Figure 3.1* Map of seventeenth-century London, illustrating how agriculture was seamlessly integrated within London. Street names that still persist today indicated what types of food were available for sale, as shown by the selected names along the major roads. Dark red lines = buildings; green parcels = fields. (From Atlas Van der Hagen et al., 1049 B 11, Fol. 20, http://www.geheugenvannederland.nl/nl/geheugen/view?coll=ngvn&identifier=KONB01%3A134. With permission.)

is one of various urbanization trends striving to revive cities. Inner-city, mixed-use communities are at the forefront of reviving urban sustainability and resilience, so that citizens can reassert control over their food security despite the large demand to feed the masses.[35]

This chapter focuses on the climate change aspects and how agroecology may help mitigate climate change and help achieve food integrity. The clash between public and private law rise to the forefront once again, where current large-scale governance is context removed and, therefore, fails to encourage citizens to take charge of the food system that is supposed to sustain them, instead of ignorantly delegating management of neoliberally induced, anthropogenic climate change[36] to federal agencies. Looking back at Figure 1.2, this discussion fits into the top left quadrant, where industrialization and urbanization contribute to pollution and, thereby, to climate change. Here, the links between GMOs, urban agriculture, and climate change lie in the standardization of feeding cities, the implied trivialization of relinquishing control of sourcing one's food by outsourcing its production, and the consequential reduction of market barriers by virtue of private law protectionism to bring said food back to urban consumers. Or, from a different perspective, food dependence is created in cities and correlates with the displacement of agroecology in urban areas because public law fails to protect agroecology more adamantly. This relationship, in turn, feeds into the right half of the cycle depicted in Figure 1.2, where the food system is streamlined and centralized.

From an urban agriculture standpoint, as Professor Wiskerke from Wageningen University in the Netherland explains,[37] food safety and food security overlap in the areas of climate change and poverty,[38] which links to food dependence and weakens food integrity. The following further contextualizes the importance of these observations in urban areas, where most of the food is consumed, the demand for GMO-based products is the highest, and where substantial pollution originates that contributes to climate change.

Thus, zooming in on the top left quadrant of Figure 2.1, an intricate relationship between food integrity and its components, agroecology, urbanization, and climate change surfaces. Specifically, with increasing

---

[35] Wiskerke, *supra* note 2, at 131.

[36] Anthropogenic climate change is defined as human-induced "widespread change detected in temperature observations of the surface,... free atmosphere ... and ocean, together with consistent evidence of change in other parts of the climate system, ... [which] strengthens the conclusion that greenhouse gas forcing is the dominant cause of warming during the past several decades." IPCC, IPCC Fourth Assessment Report at 9.7. Combining Evidence of Anthropogenic Climate Change (2007), https://www.ipcc.ch/publications_and_data/ar4/wg1/en/ch9s9-7.html (last accessed May 30, 2017).

[37] Wiskerke, *supra* note 2 (internal citations omitted).

[38] Id.

urbanization, agricultural connections of cities aggravate global warming, both by emitting heat through the urban heat island effect (UHIE) and through heavy, fossil-fuel, energy-dependent practices. Supporting and developing urban independence would reverse the process and help mitigate climate change, while simultaneously improving food integrity in cities (see Figure 3.1).

## 3.2.2    Metropolitan climate change contributions and food integrity

Food integrity and climate change are reciprocally linked: When climate change negatively affects agricultural production, food integrity suffers. Conversely, food security, a component of food integrity, often depends on pesticide-, fertilizer-, and fossil fuel-intensive conventional production that largely contributes to greenhouse gas (GHG) emissions, which, in turn, cause climate change.[39] These intertwined relationships are climate change feedback loops, "the equivalent of a vicious or virtuous circle—something that accelerates or decelerates a warming trend"[40] (see Figure 3.2). Hence, "[a] positive feedback accelerates a temperature rise, whereas a negative feedback decelerates it."[41] Applied to food security, climate change can threaten stable agricultural yields. Although long-term average temperature increases can lead to extended growing seasons and more rainfall in some parts of the world, most regions will suffer from extreme droughts and floods, which cause declines in agricultural production and food shortages.[42] Logically, the Food and Agriculture Organization (FAO) of the United Nations warns that climate change "directly affects all four dimensions of security, food availability, food accessibility, food utilization, and food systems stability."[43] In sum, cities contribute to anthropogenic climate change that, in turn, threatens food security and vice versa, thereby forming a positive feedback loop (see Figures 1.2 and 3.2).

Through city greening, such as the installation and planting of green roofs, citizens can mitigate climate change, where air pollution in large metropolitan areas is a major climate change culprit. In 2013 alone, US GHG emissions were 6673 million metric tons of $CO2$ equivalents, amounting to an average annual emission rate increase of 0.3%

---

[39] Id.

[40] What are Climate Change Feedback Loops?, *The Guardian* (undated), http://www.the-guardian.com/environment/2011/jan/05/climate-change-feedback-loops.

[41] Id.

[42] Wiskerke, *supra* note 2 (internal citations omitted).

[43] FAO, Climate Change and Food Security (Rome 2008), http://www.fao.org/forestry/15538-079b31d45081fe9c3dbc6ff34de4807e4.pdf.

Fish Freeze hill and Little Catching located north of London Bridge

Chick lane, Cow lane, Smith field

Cheap Side Poultry and Corn hill, Bread street

Red Crops street, White Crops street, and Crab street

*Figure* 3.2  Magnified excerpts from Figure 3.1.

since 1990.[44]   Correspondingly, the World Health Organization (WHO) "estimates that the warming and precipitation trends due to anthropogenic climate change of the past 30 years already claim over 150,000 lives annually,"[45] especially where the UHIE is greatest. Predictions based on the 1995 Chicago and 2003 Europe heat waves anticipate 25%–31% higher frequencies and 72%–76% longer heat waves by 2090,[46] when urban populations will likely spike while climate change probably imperils food safety (see Section 3.3.2).

Thus, Wiskerke explains that, "vegetation can be an important component of pollution control strategies in dense urban areas,"[47] where "the prevalence of air pollution in cities worsens due to the disappearance of the urban green,"[48] contributing to the heat island effect. These urban environmental health challenges are aggravated by climate change, notes Wiskerke, and "intensif[y] the energy problem of cities, deteriorate comfort conditions, put in danger the vulnerable population and amplify the pollution problems."[49] He concludes that "green roofs can play an important role in mitigating urban heat islands and, hence, in reducing the urban environmental health problems resulting from climate change."[50] Therefore, green roofs can reverse the air-pollution-urban-heat-island-effect feedback loop that challenges urban environmental health.

Agriculture's inefficiencies further complicate the relationship between urban population growth and anthropogenic climate change because much of what is being produced is wasted.[51] In the food sector, cities are generally inefficient vortices that suck resources in and spit massive amounts of waste out. The EPA reported that in 2012 alone, over 65 million tons of municipal solid waste was recovered through recycling, including packaging and waste, and over 21 million tons through composting.[52] Additionally, 29 million tons was burned for energy recovery, a whopping 2.9 pounds of discarded waste going to landfills or toxically combusted waste per person per day.[53] Moreover, as Steel notes, "[n]ineteen million hectares of rainforest are lost every year to create new arable land," and "an equivalent amount of existing arables [are lost] to salinization and

---

[44] EPA, Inventory of U.S. Greenhouse Gas Emissions and Sinks: 1990–2013 at ES-4, https://www.epa.gov/ghgemissions/inventory-us-greenhouse-gas-emissions-and-sinks.

[45] Jonathan A. Patz et al., Impact of regional climate change on human health, *Nature* 438, 310–317, 310 (Nov. 17, 2005).

[46] Id.

[47] Wiskerke, *supra* note 2 (internal citations omitted).

[48] Id.

[49] Id.

[50] Id.

[51] FAO, *supra* note 17.

[52] EPA, *Municipal Solid Waste Generation, Recycling, and Disposal in the United States* 1–14, 2 (2012), https://www.epa.gov/sites/production/files/2015-09/documents/2012_msw_fs.pdf.

[53] Id.

erosion."[54] Thus, green spaces and heat-reflective foliage are lost at alarming rates, which furthers global warming and endangers agricultural yields. As long as agriculture remains fossil fuel intensive but inefficient, where "[i]t takes about 10 calories to produce every calorie of food that we consume in the West,"[55] the future looks bleak. Returning some green spaces to urban environments can, however, disrupt some of these positive feedback loops, causing environmental deterioration and global warming, which contribute to climate change. Thus, Steel concludes, "even though there is food that we are producing at great cost, we don't actually value it. Half the food produced in the USA is currently thrown away. And to end all of this,… [a] billion of us are obese, while a further billion starve."[56] This data proves that urban food security is a ripe point to affect change. As such, introducing green roofs to cities may be a fruitful point of attack that can be supported through public laws to achieve urban food integrity.

Urban agriculture is a powerful way to tackle these problems because it reduces food miles by insourcing at least some food production; increases urban foliage, which helps to offset some of the UHIE; contributes to food production and greens cities; creates communities and employment; and brings people back in touch with their food. Another side-benefit may comprise the reduction of food miles and the shortening of supply chains, which could strengthen agroecology in the long run by decentralizing the food system at large. This promising ripple effect of bringing agriculture back to cities could, in the best-case scenario, ultimately change people's behavior toward a more climate-friendly system altogether.[57] Other benefits of urban agriculture include supplying fresh produce to low-income residents in urban food deserts, increased biodiversity, improved air quality, and storm water management.[58]

However, it is unlikely that there are sufficient vacant lots, community gardens, or potted balcony plants to offset urban pollution through urban agriculture. The built surfaces, covered in concrete, are essentially ecological losses and are unlikely to be restored in the face of growing global populations; people continue to need food and housing, places to work, park, shop, and grow their economies. Therefore, growing food on rooftops can potentially reclaim the lost space and provide a variety of environmental benefits.

---

[54] Steel, *supra* note 1.

[55] Id.

[56] Id.

[57] Susanne A. Heckler, A right to farm in the city: Providing a legal framework for legitimizing urban farming in American cities, *Val. U. L. Rev.* 47, 217, 223–227 (2012) (internal citations omitted).

[58] Matthew R. Dawson, Perennial cities: Applying principles of adaptive law to create a sustainable and resilient system of urban agriculture, *U. Louisville L. Rev.* 53, 301, 303 (2015) (internal citations omitted).

## 3.2.3   Combining urban agriculture with city greening

A delightful win-win combination of urban agriculture and city greening are rooftop farms, an augmented version of green roofs that may contribute to achieving food integrity. According to meteorological simulations, green roofs can cut the urban global warming contributions in half[59] and could reduce urban carbon emissions dramatically.[60] In fact, Michigan State University researchers "found that a square meter of vegetation captures 375 grams of carbon, which suggests greening Detroit's rooftops could remove as much carbon from the atmosphere as taking 10,000 mid-sized SUVs and trucks off the road for a year."[61] Finally, green roofs buffer storm water, sequester carbon, and improve water quality[62] in flood-prone regions.[63]

Installing URFs or retrofitting roofs to accommodate them would also bring soil back to urban landscapes, which would complement the climate change mitigation effects of increased foliage. As Diana Donlon from the Center for Food Safety explains,

> soil is the largest "sink"—or area of storage—where additional carbon would actually be extremely beneficial. Currently our cultivated soils globally have lost 50%–70% of their original carbon content. This means we have a tremendous opportunity to put carbon back into the soil where it creates positive feedback loops, making healthy soil a systemic solution to multiple problems including food and water security. Not only is rebuilding soil carbon entirely possible, … it is without risk.[64]

Not surprisingly, regaining citizen control over food security has become a revolutionary food movement[65] rich in opportunities for improvement.

---

[59] M. Georgescu, M. Moustaoui, A. Mahalov, and J. Dudhia, Summer-time climate impacts of projected megapolitan expansion in Arizona, *Nature Climate Change* 3, 37–41 (2013).

[60] Catherine Malina, Up on the roof: Implementing local government policies to promote and achieve the environmental, social, and economic benefits of green roof technology, *Geo. Int'l Envtl. L. Rev.* 23, 437, 442 (2011).

[61] Id. at 444 (internal citations omitted).

[62] See generally, Green Roofs, *2014 Annual Green Roof Industry Survey* (May 2015).

[63] Pardeep Pall et al., Anthropogenic greenhouse gas contribution to flood risk in England and Wales in autumn 2000, *Nature* 470, 382–386, 383 (Feb. 17, 2011).

[64] Diana Donlon, *Soil and Carbon*, Center for Food Safety 1–7, 2 (2015), http://www.centerfor-foodsafety.org/reports/3846/soil-and-carbon-soil-solutions-to-climate-problems.

[65] Frances Moore Lappé, The Food Movement: Its Power and Possibilities, *The Nation* (Sept. 14, 2011), http://www.thenation.com/article/food-movement-its-power-and-possibilities/.

With locally sourced organic and sustainable foods on demand, more and more people are reaching for the shovel to plant their own food, making URFs quite attractive. In fact, "15 percent of the world's food is now grown in urban areas" and, according to the FAO, "urban farms already supply food to about 700 million residents of cities, representing about a quarter of the world's urban population. By 2030, 60 percent of people in developing countries will likely live in cities."[66] A third of American households, about 41 million, already garden, which is an increase of 14% in 2009.[67] Frances Moore Lappé, a thought leader in the food movement, highlights the importance of citizens' reassertion of solutions for the opaque food system: "With an 'eco-mind' we can see through the productivist fixation that inexorably concentrates power, generating scarcity for some, no matter how much we produce."[68] If regulation and government incentives could keep up, this movement would have great potential to effect change in returning food security, food safety, and food sovereignty to urban areas while mitigating climate change. All of these efforts combine into a path toward improved food integrity in urban settings.

### 3.2.4   (W)Holistic[69] upregulation of urban agriculture

Urban food security is poorly governed and growing populations and complications are surpassing federal regulatory capacity. The US government, for instance, regulates food security through a murky cluster of laws and agency rules that are within the purview of the much-criticized "overlapping and fragmented delegations that require agencies to 'share regulatory space'" in a pervasive, dysfunctional, and stubborn food system,[70] explains Lisa Heinzerling, Georgetown Professor of Law. Consequently, bureaucratic inefficiencies, discordant policy, and non-accountability[71] jeopardize food security because these agencies enforce sometimes overlapping and interparticle statutory provisions, which are further complicated by state legislation to an obscure agglomeration of serious regulatory shortcomings.[72]

The regulatory shortcomings of the current food system and the many benefits of URFs to contribute to food security and climate change

---

[66] Food Tank, 28 Inspiring Urban Agriculture Projects (Jul. 20, 2015), http://foodtank.com/news/2015/07/urban-farms-and-gardens-are-feeding-cities-around-the-world.

[67] Id.

[68] Lappé, *supra* note 55.

[69] The term "(w)hole" appears in T. Colin Campbell and Howard Jacobson, *Whole: Rethinking the Science of Nutrition* (May 6, 2014).

[70] Rory Freeman and Jim Rossi, Agency Coordination in Shared Regulatory Space, 125 *Harv. L. Rev.* 1131, 1133 (2012).

[71] Id.

[72] See generally, Renée Johnson, The federal food safety system: A primer, *Congressional Research Service* 1–18, 14–15 (Jan. 17, 2014).

mitigation raise the question of how URFs should be promoted by local governments. There are roughly three types of regulatory schemes that URFs would fall into: (1) direct regulation, (2) collaborative governance, and (3) voluntary or market-driven governance.[73] All of these tools are highly context specific,[74] meaning that they provide a palette of context-adaptable legal tools.[75] As Dr. van der Heijden, associate professor at the University of Amsterdam, explains, the best the law can do, therefore, is distill a series of urban resilience design principles targeted to their specific contexts and problems, with a strong theoretical and empirical base.[76] All of these approaches are examples of how public law may complement private law in improving the food system. Simply put, where private law protects the industry's rights in profitable production, public law may protect consumers' and environmental interests to maintain and improve food integrity in the long run. The following examines how local governments could apply these principles to pro-URF governance.

First, through direct regulation with traditional regulatory tools, such as statutes, subsidies, tax credits, or urban gardening restrictions,[77] municipal governments could incentivize URF. These direct regulations follow three standards: prescriptive, performance-based, or goal-oriented standards.[78] Prescriptive standards rigidly try to prevent harmful events, but they lack flexibility and may stifle innovation.[79] Performance-based standards, such as energy efficiency specification, overcome some of the shortcomings of prescriptive standards, but leave it to those regulated to find a way to comply most efficiently.[80] Conversely, target or goal-oriented standards seek to prevent harmful events by directly linking outcomes with regulatory goals, leaving it fully to those regulated to achieve compliance.[81]

In Seattle, Washington, for example, direct regulations under a comprehensive plan pursuant to Washington's Growth Management Act required Seattle to manage population growth and helped the city to reduce barriers to urban agriculture, and thereby URFs.[82] For instance, incentive zoning and density bonuses encourage livability and

---

[73] van der Heijden, *Governance for Urban Sustainability and Resilience.* Edward Elgar Publishing (2014); ProQuest ebrary.

[74] Id. at 124.

[75] Id.

[76] Id.

[77] Id.

[78] Id. at 33 (internal citations omitted).

[79] Heijden, *supra* note 57 at 33–34.

[80] Id. at 34 (internal citations omitted).

[81] Id.

[82] Urban Agriculture in Seattle, https://assets.jhsph.edu/clf/mod_clfResource/doc/Urban%20Agriculture%20in%20Seattle%20Policy%20and%20Barriers.pdf.

sustainability.[83] They "link code flexibility, increased density and development potential with public benefits in the form of affordable housing and other amenities valued by communities,"[84] thereby setting a prescriptive standard. Seattle encourages city planning with innovative land use management techniques, including density bonuses and cluster housing.[85] Clustering "means shorter streets and fewer impervious surfaces that dump runoff into streams" and should ideally be at rural densities outside metropolitan areas.[86] Such cluster zoning "allows new development on one portion of the land, leaving the remainder in agricultural or open space uses"[87] for a variety of innovative zoning techniques designed to conserve agricultural lands and encourage the agricultural economy.[88] Alternatively, incentive zoning offers a density bonus on new development in exchange for community amenities, such as granting developers additional height in exchange for providing public benefits.[89] Direct regulations have a mixed track record but can be adapted to the URF context.[90]

Second, Haijden explains that through collaborative governance tools, that is, networks, covenants, and agreements, stakeholders share regulatory authority.[91] In the food systems context, lifting some of the regulatory burden from the clustered agency oversight could provide welcome resolutions and simplifications. In other words, deregulation, privatization, and public management can take over some of the jobs that agencies are burdened with.[92] Haijden also notes that collaboration may improve regulatory efficiency because non-governmental actors have better knowledge of their day-to-day behavior than governments and can, therefore, cater to local needs.[93] Such collaboration empowers those governed to be responsible for co-regulating, thereby improving accountability and transparency.[94] Thus, collaborative city regulators unite efforts and merge diverse interests to ensure cohesion and efficiency among various actors by enforcing laws and facilitating

---

[83] Seattle Planning Commission, Incentive Zoning in Seattle 1–10, 4 (2007), http://www.seattle.gov/Documents/Departments/SeattlePlanningCommission/IncentiveZoning/SPC_Incentive%20Zoning.pdf.

[84] Id.

[85] Wash. Rev. Code Ann. § 36.70A.090 (West).

[86] Eric Pryne, "Cluster" Concept on Rise But Not Universally Liked, *Seattle Times* (Mar. 4, 2008), http://www.seattletimes.com/seattle-news/cluster-concept-on-rise-but-not-universally-liked/.

[87] Wash. Rev. Code Ann. § 36.70A.177(2)(b) (West).

[88] Id. at § 36.70A.177(1).

[89] Heijden, *supra* note 60 at xi.

[90] N.B. Markets are not forms of regulation because they are not prescriptive or prohibitive. Markets do, however, function like regulation or even more powerfully because they have effects on systems at large.

[91] Id.

[92] Id. at 61.

[93] Id.

[94] Id. at 63.

success through incentives.[95] However, a downside of collaborative governance is that it is prone to manipulation because "the voices of weaker stakeholders are rarely included in the outcome of such processes" and, are thus excluded from the collective decision-making process.[96]

In Chicago, Illinois, for example, the Chicago Sustainable Backyards Program "calls on residents to help make Chicago more sustainable, one backyard at a time" as "part of the solution to issues surrounding water, waste, energy, and habitat loss."[97] The program is USDA funded,[98] and provides resources, discounts, and a network of services to facilitate backyard greening, composting, and water management.[99] On a larger scale, the program is designed to protect Lake Michigan and improve green infrastructure. Sustainable Backyards rebate incentives help improve soils, cool air, and foster a sense of community.[100]

Another example is the Fenway Farms co-op model in Boston, Massachusetts, where URFs have recently hit the big league.[101] Boston's stadium sources food from its own "in-house" garden on the Red Sox's Fenway Park. The brand new 5,000 square feet URF will produce an estimated 40,000 pounds of arugula, kale, radishes, carrots, and other fresh produce for use at the stadium's EMC club.[102] Sales will likely exceed $25,000, bringing the co-op under the Food Safety Modernization Act's (FSMA) jurisdiction. The regulatory opportunity for Boston to co-regulate with co-op stakeholders could relieve some of the Food and Drug Administration's (FDA) burden of enforcing the statute. Simultaneously, such precautionary co-op governance collaboration would facilitate food safety for an initiative that might otherwise fly under the FDA's radar until a foodborne disease outbreak would draw attention to the EMC club. This is a prime example of how private and public law affect urban green roofs that are used for agricultural purposes. More projects like Fenway Farms could help cities achieve food integrity.

Third, voluntary and market-driven governance, such as best-of-class benchmarking and certification, tripartite financing, green leasing,

---

[95] Id.

[96] Id. at 66.

[97] City of Chicago, Creating Sustainable Backyards Gets Easier For Chicagoans (Apr. 22, 2013), http://www.cityofchicago.org/city/en/depts/cdot/provdrs/conservation_outreachgreenprograms/news/2013/apr/creating_sustainablebackyardsgetseasier-forchicagoans.html.

[98] Id.

[99] See generally, sustainablebackyards.org.

[100] City of Chicago, *supra* note 83.

[101] Food Tank, Urban Farming Hits the Big Leagues (Nov. 29, 2015), http://foodtank.com/news/2015/11/urban-farming-hits-the-big-leagues.

[102] Green City Growers, The Fenway Farms System (undated video), http://greencitygrowers.com/fenway-farms/fenway-farms-installation/.

contests and challenges, and sustainable procurement,[103] echo direct regulatory interventions in a deceiving *laissez-faire* approach.[104] This self-governance fits the shift "from government to governance," similar to collective regulation (of the non-legal sort), but differs in that participating businesses or citizens take more cost-effective action than the government enforcement of laws to ensure that the collective goals are met.[105] For instance, the ClimateWise program in Fort Collins, Colorado, has supported 370 businesses in over 8570 projects,[106] saved over $92 million, and reduced 1.3 million metric tons of carbon dioxide equivalent since 2000, by providing a menu of best management practices and local resources to support GHG reduction goals.[107] It follows that lessons for future regulatory approaches supporting URFs support a holistic approach that tackles urban problems collectively, that enforces and incentivize compliance, and supports further innovation to tackle emerging challenges for city greening and urban food integrity.[108]

All three forms of governance take a holistic approach to upregulate food security, urban population growth and resilience, environmental conservation, and climate change mitigation—crucial steps toward achieving food integrity. The effect of these types of governance would be reversing positive feedback loops where food insecurity and climate change feed into a vicious circle (see Figure 3.2). Similarly, these incentives create points of attack for strategic disruptions of the cycle illustrated in Figure 1.2. Thus, if agencies overlook the potential of URFs, the aforementioned approaches could be adapted to URF's needs in any major US city and relieve the burden on agencies.

Augmented mitigation measures through rooftop greening and urban farming can improve resilience and address the three priorities in climate change adaptation of developed areas:

1. Reducing the costs and disruptions from extreme weather events like storms, floods, and heat waves
2. Stimulating public adaptation initiatives
3. Developing market structures with high adaptive capacity[109]

---

[103] Heijden, *supra* note 60.

[104] Id. at 88.

[105] Id.

[106] Fort Collins Infographic, http://www.fcgov.com/climatewise/pdf/cw_2014operationsreport.pdf.

[107] Fort Collins, Climate Wise, http://www.fcgov.com/climatewise/index.php.

[108] Adapted from Heijden, *supra* note 60 at 138–139.

[109] Nicholas Stern, Adaptation in the developed world at 471, in *The Economics of Climate Change*. Cambridge University Press (2007).

More importantly, however, insourcing food production may fix far-reaching problems of the current food system by, for example, eradicating environmentally harmful GMO-dependent monocultures and replacing them with sustainable URFs. The high urban population density provides a consistent customer base and an inherent task force to carry the urban farming models, especially on roofs, which provide "arable land" outside the real estate competition. Therefore, context-specific holistic upregulation is needed in the sense that existing zoning and local ordinances should not only encourage and incentivize, but also reward local food procurement from within the cities to reduce legal barriers and methods of insourcing to feed cities.

Although cities are the problem, they can also be the solution.[110] The key is to bring agriculture back into cities. Benefits of urban agriculture all support URFs, as well as improved food security and cost internalization, and the economics of climate change. Improving food security in large cities by insourcing food production and bringing agriculture back to the cities would, at least partially, solve a host of problems, such as strengthening the local food movement to reduce food miles and encouraging city greening to mitigate climate change. It is an inside-out approach the closes many gaps of the current food system, specifically the resource wastefulness and absurd inefficiencies of megapolitan areas. Food can powerfully shape the world, and URF regulations could be a supportive tool toward food integrity.

## 3.3 Creating infrastructure to circumvent GMOs: Toward environmental resilience through urban rooftop farming

Considering that the UHIE basically gives cities a fever from a climate change perspective, the question arises where points of attack are available to tackle the problem. Again, the reliance on GMOs in a system where food production is outsourced, feeds into the dangerous cycle illustrated in Chapter 1 (see Figure 1.2). It follows that halting the proliferation of GMOs and the associated food dependence in urban environments lends itself to finding a solution. This section explains how zoning laws can address the proliferation of GMOs from the urban climate change perspective, with the goal to increase urban agriculture and improve access to non-GMO foods in cities, thereby strengthening food integrity. In short, bringing agriculture to cities shortens supply chains for fresh produce, thereby creating an infrastructure to circumvent GMOs and promote agroecology in the urban setting. When city dwellers grow organic produce on their

---

[110] Grimm et al., Global change and the ecology of cities, *Science* 319, 5864, 756–760 (2008).

roofs and in areas with high population density, they automatically have a market for it, which makes them land stewards (strengthening food sovereignty's urban counterpart), potentially providing fast-food deserts with nutrient-dense foods (improving food security) and simplifying oversight of this local agriculture (improving food safety).

Zoning Against Climate Change with Green-Roof Legislation[111] offers promising solutions to many factors leading to urban outsourcing and the proliferation of GMOs. At its core lies the problem of city greening. Correspondingly, the space in green cities is on roofs. Policy analysts embrace this solution. According to Elizabeth Kucinich, policy analyst and environmental advocate,[112] green roofs provide many benefits: "increased efficiency for buildings, heat sinks, an opportunity to create natural habitat, much-needed urban recreational space, water catchment, air particulate capture, good jobs—for construction and horticulture, even food production and the potential for carbon sequestration."[113] Thus, green roofs, if supported through public laws at the local level, may help to achieve food integrity by reducing some of the risks of a GMO-centric food system and by encouraging agroecology.

Overall, green roofs offer a simple but often overlooked solution to the complex problem of anthropogenic climate change.[114] This chapter proposes green-roof zoning laws to mitigate heat pollution from urban microclimates known as the UHIE, the effect of urban heat absorption and radiation on impervious surfaces, such as concrete, asphalt, and brick.[115] In the United States and the EU, rampant federal inaction misses opportunities to rise against the local environmental challenges of urbanization, heat pollution, and climate change. In Germany, for example, Stuttgart[116] has become the most polluted city exceeding several emission

---

[111] An abbreviated version of this section has been published in the *Vermont Law Journal* Blog.

[112] Interview with Elizabeth Kucinich (Oct. 6, 2015), on file with author.

[113] Id.

[114] For the purpose of this chapter, climate change includes global warming through human-caused heat pollution. Climate change, under Article 1 of the United Nations Framework Convention on Climate Change (UNFCCC) [is] define[d] as "a change of climate which is attributed directly or indirectly to human activity that alters the composition of the global atmosphere and which is in addition to natural climate variability observed over comparable time periods...." The UNFCCC thus makes a distinction between "climate change" attributable to human activities altering the atmospheric composition, and "climate variability" attributable to natural causes. World Meteorological Organization, What is Climate Change?, http://www.wmo.int/pages/prog/wcp/ccl/faqs.php (last accessed Oct. 18, 2015).

[115] Catherine Malina, Up on the roof: Implementing local government policies to promote and achieve the environmental, social, and economic benefits of green roof technology, *Geo. Int'l Envtl. L. Rev.* 23, 437, 442 (2011).

[116] *Süddeutsche Zeitung* (Southern German newspaper), Stuttgart hat die schlechteste Luft [Stuttgart has the worst air] (April 24, 2015) (translation by author), http://www.sueddeutsche.de/wissen/feinstaubbelastung-stuttgart-hat-die-schlechteste-luft-1.2450361 (last accessed Oct. 18, 2015).

thresholds.[117] As the hub of major car producers Audi and Mercedes-Benz, Stuttgart is "the cradle of the automobile." It resembles Detroit in the United States, but for the major distinguishing factor that Stuttgart has prevented a mass exodus and post-industrialization impoverishment. Since the 1970s, through zoning ordinances, Stuttgart has recovered environmental balance and has become an exemplary forerunner in city greening. Megapolitan areas in the United States could draw from Stuttgart's zoning models to improve urban habitability, public health, and sustainability. This section explores how zoning laws in the United States can be modeled after Stuttgart's example to reduce the UHIE.

Notably, however, this section does *not* suggest that a legal analysis from one culture can be transferred to another culture.[118] Law is a part of culture and, unlike natural science phenomena, legal tools are not the same regardless of the culture.[119] This section merely compares how transplantation of Germany's city-greening initiative in Stuttgart might help similar metropolitan areas in the United States, considering the cultural and legal context within the scope of this book.

One method of city greening is roof greening, planting vegetation atop buildings. Green roofs provide shading and moderate air temperatures through plant evapotranspiration and offset some of the UHIE.[120] However, legislative action overlooks green roofs as important tools to mitigate climate change. Municipal legislators may amend zoning laws to incentivize sustainable city planning based on Stuttgart's example. The important and novel approach in this chapter, inspired by Stuttgart's climate mapping, suggests green-roof zoning laws to reduce the UHIE. Initially, Section 3.3.1 explains the connections between the UHIE, climate change, and green roofs. Subsequently, Section 3.3.2 introduces the laws underlying Stuttgart's successful city greening and discusses the shortcomings of existing US green-roof policy. Finally, Section 3.3.3 introduces a model US zoning approach to reduce the UHIE.

## 3.3.1 Targeting climate change with zoning laws

The UHIE affects public health and contributes to climate change. According to the US Environmental Protection Agency (EPA), the UHIE occurs in metropolitan areas and "can affect communities by increasing summertime peak energy demand, air conditioning costs, air pollution and greenhouse gas emissions, heat-related illness, ... mortality, and

---

[117] Id.
[118] Correspondence with Professor Junker.
[119] Id.
[120] Id.

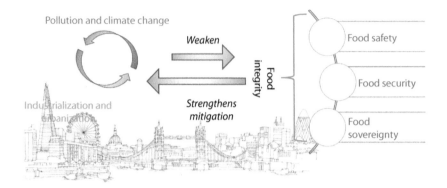

*Figure 3.3* Relationship between (1) pollution and climate change, (2) industrialization and urbanization, and (3) food integrity. This figure illustrates the relationships between pollution and climate change, which make up a self-perpetuating cycle with industrialization and urbanization, feeding into each other. This cycle weakens the chances of achieving food integrity (gray arrow pointing right). Considering the reverse, food integrity, with its components food safety, security, and sovereignty, has the potential to strengthen climate change mitigation, reduce industrialization and pollution, and use urbanization toward a more positive outcome, which, in turn, may help to strengthen food integrity.

water quality."[121] Correspondingly, the WHO "estimates that the warming and precipitation trends due to anthropogenic climate change of the past 30 years already claim over 150,000 lives annually,"[122] especially where the UHIE is greatest. In fact, the EPA estimates that "a city with 1 million people or more can be 1.8–5.4°F (1–3°C) warmer than its surroundings"[123] during the day and 22°F (12°C) in the evening (see Figures 3.3 through 3.5).[124] Predictions based on the 1995 Chicago and 2003 Europe heat waves anticipate 25%–31% higher frequencies and 72%–76% longer heat waves by 2090.[125] Studies in the largest metropolitan areas in the United States also confirm the links between mortality,[126] the UHIE, and climate change.

---

[121] EPA, Heat Island Effect, http://www2.epa.gov/heat-islands (last accessed Sept. 30, 2015). *See also* J. Lelieveld et al., The contribution of outdoor air pollution sources to premature mortality on a global scale, *Nature* 525, 367–370 (Sept. 17, 2015). In fact, "[o]n sunny summer days, city roofs and pavement, which are dry and exposed, may be 50–90°F hotter than natural surfaces, which are shaded or moist." Id.

[122] Jonathan A. Patz et al., Impact of regional climate change on human health, *Nature* 438, 310–317, 310 (Nov. 17, 2005).

[123] EPA, *supra* note 8.

[124] Id.

[125] Patz, *supra* note 9, at 314.

[126] Frank Curriero et al., Temperature and mortality in 11 cities of the Eastern United States, *Am. J. Epidemiol.* 155 (2002), 80–87, 81.

*Figure 3.4* Satellite data image of night light indicators of urban heat islands: Europe, Northern Africa, and Middle East. True climate change is visualized in metropolitan areas (white), where the temperature anomalies are the largest. (From NASA, Satellites Shed Light on a Warmer World, Nov. 5, 2001, https://www.giss.nasa.gov/research/news/20011105/. With permission.)

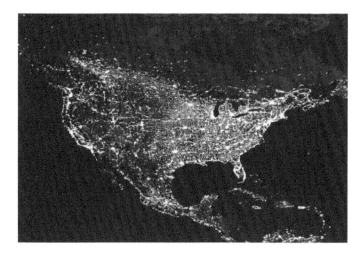

*Figure 3.5* Satellite data image of night light indicators of urban heat islands: North and Central America. True climate change is visualized in metropolitan areas (white), where the temperature anomalies are the largest. (From NASA, Satellites Shed Light on a Warmer World, Nov. 5, 2001, https://www.giss.nasa.gov/research/news/20011105/. With permission.)

According to meteorological simulations, green roofs can cut the urban global warming contributions by half[127] and could reduce urban carbon emissions dramatically.[128] A recent report published by the National Academy of Science explains that,

> continued conversion of existing lands to urban landscapes has the potential to drive significant local and regional climate change, compounding global warming. At the same time, how cities choose to expand and develop will be critical to defining how successful society will be in adapting to global change. Because cities are, in a real sense, fundamental units of both climate change adaptation and mitigation, development choices in the coming century will lead to either significant exacerbation or significant reduction in the impacts of global change.[129]

Additionally, Michigan State University researchers "found that a square meter of vegetation captures 375 grams of carbon, which suggests greening Detroit's rooftops could remove as much carbon from the atmosphere as taking 10,000 mid-sized SUVs and trucks off the road for a year."[130] Finally, green roofs buffer some storm water, sequester carbon, and improve water quality[131] in flood-prone regions.[132]

Although international and national programs address the UHIE, zoning is strategically better positioned to tackle urban microclimates, which accumulate and contribute about 4°C to global warming. Notably, global urbanization[133] facilitates UHIE accumulation, thereby driving climate change. According to the United Nations Environment

---

[127] M. Georgescu, M. Moustaoui, A. Mahalov, and J. Dudhia, Summer-time climate impacts of projected megapolitan expansion in Arizona, *Nature Climate Change* 3, 37–41 (2013) (Aug. 12, 2012), doi:10.1038/nclimate1656.

[128] Malina, *supra* note 4, at 444 (internal citations omitted).

[129] Matei Georgescu, Philip E. Morefield, Britta G. Bierwagen, and Christopher P. Weaver, Urban adaptation can roll back warming of emerging megapolitan regions, 111 *PNAS* 8 (2014), 2909–2914, 2909; published ahead of print February 10, 2014 (internal citations omitted).

[130] Malina, *supra* note 4, at 444 (internal citations omitted).

[131] See generally, Green Roofs, *2014 Annual Green Roof Industry Survey* (May 2015).

[132] Pardeep Pall et al., Anthropogenic greenhouse gas contribution to flood risk in England and Wales in autumn 2000, 470 *Nature* 382–386, 383 (Feb. 17, 2011).

[133] By 2050, the United Nations, Department of Economic and Social Affairs, projects that 66% of the world population will live in urban areas, a nearly 10% increase to current data. See United Nations, Department of Economic and Social Affairs, *World Urbanization Prospects: The 2014 Revision* (2014) 1–27, at 1, http://esa.un.org/unpd/wup/Highlights/WUP2014-Highlights.pdf (last accessed Oct. 1, 2015).

Programme, "30%–40% of all primary energy is used in buildings,"[134] contributing greatly to fossil fuel use and climate change, especially in the United States, where 82% of people live in cities.[135] Cutting these figures in half through green roofs would mitigate climate change significantly.[136]

Correspondingly, in the United States, land use laws aim at environmental protection after urban development,[137] but no specific regulation addresses the UHIE. The EPA maintains a Heat Island Reduction Program,[138] which merely provides non-binding guidance for local action.[139] Through green roofs, the overall energy consumption of the building is lowered,[140] and green roofs "provide a beneficial environmental modification that reduced building energy needs and protects against two current public health stressors: high summertime heat and ground-level ozone...."[141] It follows that positive zoning for roof greening may spark legal action in the United States. Although business and environmental cases for green roofs have been established, and despite the surge in green or high-energy performance building construction, no streamlined approaches of government intervention have emerged to support and incentivize green roofs.[142] This section suggests green-roof zoning to tap the benefits for US cities.

[134] Carl J. Circo, Using mandates and incentives to promote sustainable construction and green building projects in the private sector: A call for more state land use policy initiatives, *Penn St. L. Rev.* 112, 731, 733 (2008) (citing Pekka Huovila et al., U.N. Env't Programme, *Buildings and Climate Change: Status, Challenges and Opportunities* v [2007]).

[135] Id.

[136] Similarly, the Intergovernmental Panel on Climate Change (IPCC) identifies green roofs and city greening as important measures to mitigate the UHIE. See IPCC, Urban Heat Islands and Land Use Effects, https://www.ipcc.ch/publications_and_data/ar4/wg1/en/ch3s3-2-2-2.html; The UN Earth Summit, called "for standards to be organized around the objective of supporting ongoing economic development while preserving the earth's resources for future generations." See United Nations Conference on Environment and Development (UNCED), Rio de Janeiro, 3–14 June 1992, http://www.un.org/geninfo/bp/enviro.html.

[137] Circo, *supra* note 21 (internal citations omitted).

[138] EPA, Heat Island Effect, http://www.epa.gov/heatisld/ (last accessed Sept. 30, 2015).

[139] EPA, What EPA is Doing to Reduce Heat Islands, http://www2.epa.gov/heat-islands/what-epa-doing-reduce-heat-islands (last accessed Sept. 30, 2015) (noting that "EPA's Heat Island Reduction Program (HIRP) focuses on translating urban heat island research results into outreach materials, tools, and guidance to provide communities with information needed to develop urban heat island projects, programs, and policies.").

[140] Id. at 736–737 (internal citations omitted).

[141] Columbia University Center for Climate Systems Research & NASA Goddard Institute for Space Studies, Green Roofs in the New York Metropolitan Region: Research Report 2 (Cynthia Rosenzweig, Stuart Gaffin, and Lily Parshall, Eds., 2006), https://pubs.giss.nasa.gov/abs/ro05800e.html (last accessed Sept. 30, 2015).

[142] Circo, *supra* note 21, at 732.

### 3.3.2 *Comparing German laws favoring green roofs and lessons from the Stuttgart case study*

Considering how much benefit green roofs could bring to US cities, it is important to explore how other cities implemented city greening. Turning the perspective to a city in Germany and comparing the incentives there to those that might work in the United States, the following case study informs a comparison of EU-US strategies to tackle the urbanization that fuels the proliferation of GMOs and food dependence, as noted in Section 3.2. Additionally, this case study illustrates how much-needed public law may protect agroecology and pave the way toward improved food integrity by complementing the protectionist approach of private law in the US food system.

In Stuttgart, public laws, namely the Federal Nature Conservation Act (BNatSchG),[143] the Nature Conservation Act of the state of Baden-Württemberg (NatSchG),[144] and the federal German building code (BauGB),[145] govern the preservation of green spaces, including green roofs. The overarching general principle of the federal BNatSCHG is "to permanently safeguard (1) biological diversity, (2) the performance and functioning of the natural balance, including the ability of natural resources to regenerate and lend themselves to sustainable use, and (3) the diversity, characteristic features and beauty of nature and landscape...."[146] Additionally, §3(4) prioritizes the protection of "the air and the climate ... via measures of nature conservation and landscape management."[147] On a state level, the NatSchG[148] promulgates a wide array of nature conservation mechanisms and emphasizes the need for urban planning for climate

[143] Act on Nature Conservation and Landscape Management (Federal Nature Conservation Act—BNatSchG) of 29 July 2009, unofficial translation, http://www.bmub.bund.de/fileadmin/Daten_BMU/Download_PDF/Naturschutz/bnatschg_en_bf.pdf (last accessed Oct. 2, 2015).

[144] Kasmirczak, A. and Carter, J., Stuttgart: Combating heat island and poor air quality with green aeration corridors, (2010) at 3 Adaptation to climate change using green and blue infrastructure, Univ. of Manchester (internal citations omitted), http://climate-adapt.eea.europa.eu/metadata/case-studies/stuttgart-combating-the-heat-island-effect-and-poor-air-quality-with-green-ventilation-corridors.

[145] Baugesetzbuch (BauGB) (translation by author), http://www.gesetze-im-internet.de/bundesrecht/bbaug/gesamt.pdf.

[146] BNatSchG §1(1). Moreover, this "protection shall include management, development and ... restoration of nature and landscape..." Id.

[147] BNatSchG §3(4).

[148] Gesetz des Landes Baden-Württemberg zum Schutz der Natur und zur Pflege der Landschaft (Naturschutzgesetz—NatSchG) (translation by author), http://www.landesrecht-bw.de/jportal/?quelle=jlink&query=NatSchG+BW&psml=bsbawueprod.psml&max=true&aiz=true (last accessed Oct. 2, 2015).

change mitigation.[149] The revised 2004 BauGB "requir[ing] precautionary environmental protection in urban zoning and planning practices" by "facilitat[ing] air exchange in the city and enhanc(ing) cool air flow … by specific measures aimed at the maintenance and enhancement of open spaces and provision of vegetation,"[150] specifically including green roofs. This revision created a culture of city greening that extended to Stuttgart's roofs and even industrial buildings.

Accordingly, Stuttgart's Urban Climatology Office implemented these provisions by mapping "regional wind patterns, flows of cold air, air pollution concentrations, and … other relevant information required to inform planners on what to do for urban climatic optimization in new projects and retrofits."[151] With these maps, urban planners use green roofing to address social, economic, or ecological impact[152] of heat pollution and "provid[e] technical support for decision-making regarding land use planning."[153] Specific city-greening zoning based on the climate atlas maps reduces the UHIE and incentivizes city greening (Figures 3.6 and 3.7).

### 3.3.3  Fruitful comparison: US green-roof policy models

In the United States, green-roof policy insufficiently targets the UHIE and exemplifies the problem of insufficient public law protection for agroecology and, implicitly, policy favoring food integrity. Such needed policy includes pilot projects, financial incentives, and command-and-control strategies,[154] which lag behind Stuttgart's model in practicality, likelihood of success, and legislative impact to create a culture of city greening resembling the one in Stuttgart.

First, pilot projects, such as the green roof on Chicago's City Hall and San Antonio's federal courthouse,[155] illustrate how government or large commercial buildings can spearhead roof greening, but they are prohibitively expensive without local government support.[156] Moreover, market

---

[149] Kasmirczak et al., *supra* note 31 (internal citations omitted). For instance, BauGB § 1(5) requires "urban development planning … to be sustainable, integrating social, economic and ecological demands, and also assuming responsibility for future generations. Urban development plans must contribute to the creation of an environment that is fit for human beings, that protects natural resources, which contributes to climate protection…" Id.

[150] Id. at 3.

[151] Id. at 4.

[152] Peter Stott et al., Human contribution to the European heatwave of 2003, *Nature* 432, 610–613, 611 (Dec. 2, 2004).

[153] Id.

[154] Malina, *supra* note 4, at 446.

[155] Live Roof, Green Roof Centerpiece of LEED Platinum Courthouse Renovation, http://www.liveroof.com/green-roof-centerpiece-of-leed-platinum-courthouse-renovation/ (last accessed Oct. 18, 2015).

[156] Id. at 457.

*Figure 3.6* Stuttgart's climate atlas map. This classified thermal map illustrates the highest UHIE in Stuttgart. The red and orange zones have the greatest variance from normal temperatures and overlap with the most densely populated areas with the least vegetation where the UHIE is most significant. The green and blue areas show the coolest regions, which coincide with green areas and woods. (From Klimaatlas Region Stuttgart at 59 and 81 (May 2008), https://www.stadtklima-stuttgart.de/index. php?klima_klimaatlas_region.)

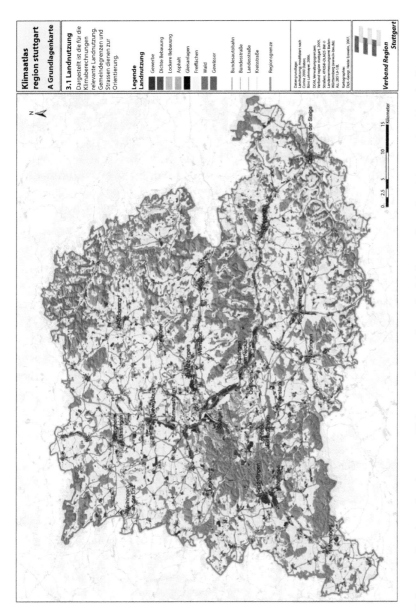

*Figure 3.7* Stuttgart's urban density map. This map shows where the highest urban densities are (pink), which are congruent with the hottest zones (red in Figure 3.6). (From Klimaatlas Region Stuttgart at 59 and 81 (May 2008), https://www.stadtklima-stuttgart. de/index.php?klima_klimaatlas_region.)

barriers and existing zoning laws often impede green-roof construction.[157] Thus, pilot projects may remain only examples and curtail setting national roof-greening trends without effective green-roof legislation.

Second, general financial incentives reduce market barriers.[158] Portland, Oregon, for example, actually used zoning in its direct financial incentive program, the "Floor Area Ratio (FAR) bonus offering developers a less restricted height-to-floor-area ratio for projects with green roofs...."[159] Although direct economic incentives, "such as grants and subsidies, ease the initial cost of green roof installations" [160] and are somewhat successful,[161] these plans generally do not create a lasting commitment and ignore the underlying reason[162] of mitigating climate change because they are private legal measures with limited reach, as opposed to public law programs. As long as investment risk is not reduced, such as through Managed Energy Services Agreements (MESA®),[163] they risk maintaining the environmentally dangerous status quo in city planning. Ideal green-roof zoning laws would create incentives for large and small developers to build green roofs on any size of buildings, whether newly constructed or retrofitted.

Third, current building regulation is opaque and inconsistent. For example, "[r]egulation strategies include performance standards, setting requirements for solar reflectivity, green space, and stormwater management" and "establish technology standards, mandating the use of green-roof technology on new development or renovation projects."[164] While regulations may be a "[p]articularly useful strategy when buildings

[157] Id.

[158] Id. at 449.

[159] Id.

[160] Malina, *supra* note 4, at 449.

[161] "As of 2007, the program had facilitated an estimated $225 million in additional private development, with more than 120 ecoroofs constructed in the city's central district. Chicago similarly permits higher density development in its central business district if more than fifty percent or 2000 square meters of roof surface area is vegetated." Id. at 452 (internal citations omitted).

[162] Alfie Kohn, Why Incentive Plans Cannot Work, *Harvard Business Review* (Sept. 1993), https://hbr.org/1993/09/why-incentive-plans-cannot-work (last accessed Oct. 1, 2015).

[163] A Managed Energy Services Agreement (MESA®) is an innovative business model created by SCIenergy that uses hidden energy savings to finance infrastructure upgrades. Under MESA®, building owners pay SCIenergy a fee equivalent to or slightly less than their historical utility bill while SCIenergy assumes responsibility for the owner's utility bills and pays for any upgrades through savings, disrupting traditional models by minimizing investment risk and project responsibility for owners. SciEnergy, MESA® (Managed Energy Services Agreement), https://www.rila.org/sustainability/RetailEnergyManagementProgram/Documents/MESA%20Primer.pdf (last accessed Oct. 19, 2015).

[164] Malina, *supra* note 4, at 453.

are subject to the development approval process and limited budgets,"[165] legislation is slow and unpopular with developers.[166]

In sum, legal tools in the United States exist to promote green roofs but local legislation could complement them to overcome current shortcomings. Common goals between green-roof zoning initiatives and local legislation could be drivers to achieve food integrity, but also wide-reaching environmental conservation and public health interests with substantial economic impact.

### 3.3.4 Rethinking zoning

This proposal relies on a city climate atlas emulating Stuttgart's model, which visualizes the areas where the zoning laws apply and makes the information easily accessible and scientifically reliable. In response to Congress' lacking city-greening incentives, the herein proposed zoning via locally targeted executive acts[167] would bridge gaps between legislation and city-greening policy.

US zoning laws may follow Stuttgart's implementation of the BNatSchG, NatSchG, and BauGB through green-roof zoning legislation. Although Stuttgart's Climate Atlas did not specifically map where green roofs should be implemented, it can be combined with the pertinent language from the governing federal and state laws to inform city planning on the ideal locations of green roofs and incentivize roof greening. By analogy, local legislative action, informed by the EPA's heat island effect program, may target microclimates to mitigate climate change[168] by tailoring US zoning laws to regional city-greening requirements.

A sample US ordinance, for example, may require that a city zones by following steps that resemble Stuttgart's successful approach of city greening:

1. Create a climate atlas to map the standardized climatic evaluations of the region.
2. Identify and prioritize the most effective regions for green roofing.

---

[165] Id. at 459.

[166] Id.

[167] Dibbs v. Hillsborough Cnty., Fla., 67 F. Supp. 3d 1340, 1352 (M.D. Fla. 2014) aff'd, No. 15-10152, 2015 WL 5449225 (11th Cir. Sept. 17, 2015).

[168] Id. at 443–444 (internal citations omitted) ("In 2006, buildings accounted for seventy-two percent of total electricity consumption in the United States, and almost forty percent of the nation's carbon dioxide emissions came from buildings. However, buildings with green roofs can consume fourteen percent less energy annually and have indoor air temperatures 4°F cooler than buildings without green roofs, as their vegetated surfaces moderate heat flow and reduce average daily energy demand for air conditioning by an estimated seventy-five percent. The 2.5-acre green roof atop the U.S. Postal Service distribution facility in midtown Manhattan has already sliced the building's annual energy bill by $30,000. Greening all of the rooftops in Chicago would save an estimated $100 million in annual energy costs and cut peak electricity demand by 720 megawatts, reducing greenhouse gas emissions by a significant margin.")

3. Subsidize, reduce market barriers, or allow for state tax credits to encourage rooftop greening and create experiment sites for evaluation and optimization.
4. Maintain a five-year program to reevaluate and update the local climate atlas to encourage continued and strategic rooftop greening and to adapt plant types to local needs (including crops for urban farming).

For the first two steps, local universities or meteorological laboratories could assist the local regulators to assess topography, development density, and air flow requirements to meet the cooling effects offsetting the UHIE. For the second and third steps, local legislation would provide the respective means to promote green roofs. Local agencies may monitor test sites to inform Steps 3 and 4 optimizing vegetation and raising city support for green roofs in five-year increments.

Ideal green-roof zoning laws would combine features already nested in municipal urban agriculture, impervious overlay zoning,[169] or tax codes. Boston Zoning Code §89(5), for example, allows both open-air and greenhouse rooftop farming "in all [l]arge-scale [c]ommercial, [i]ndustrial, and [i]nstitutional [d]istricts and subdistricts."[170] The benefits of urban green-roof farming are "boosting food access in Boston's underserved communities, providing new opportunities for local business growth, and developing knowledge and education about healthy eating." They "can be a source of fresh produce for neighborhoods, local restaurants and … an opportunity for community-supported enterprises…."[171] Additionally, green roofs slow storm water runoff[172] in mapped districts that zone for total impervious cover limits. Urban farms on rooftops, for example, oasify barren roof surfaces, thereby effectively using otherwise lost space. Applied to megapolitan heat islands, reversing the heat desertification by planting rooftop gardens, it may also help to expand urban farming. The benefits of green roofs in Boston can be easily translated to other cities and beyond summer UHIE.

---

[169] EPA, Zoning, http://water.epa.gov/polwaste/npdes/swbmp/Zoning.cfm (last accessed Oct. 18, 2015).
[170] Bos. Art. 80, Urban Agric. (Dec. 20, 2013), http://www.bostonredevelopmentauthority. org/getattachment/a573190c-9305-45a5-83b1-735c0801e73e (last accessed Oct. 17, 2015). *See also* Boston Redevelopment Authority, http://www.bostonredevelopmentauthority. org/planning/planning-initiatives/urban-agriculture-rezoning (last accessed Oct. 17, 2015).
[171] Article 89 Made Easy: Urban Agriculture Zoning for the City of Boston, http:// www.bostonredevelopmentauthority.org/getattachment/5579e854-b3c5-49e6-b910-fedaa2dd6306 (last accessed Oct. 17, 2015).
[172] Erica Oberndorfer et al., Green roofs as urban ecosystems: Ecological structures, functions, and services, *BioScience* 57, 10 (November 2007), 823–833.

Drawing from successful programs in a utilitarian comparative method might yield a spectrum of tools to adapt the green-roofing solution to various cities around the world. Comprehensive, citywide green-roof acts from Toronto[173] or New York's tax abatement[174] should proliferate green roofs to offset the respective city's heat pollution. Minneapolis, for instance, incentivizes green roofing indirectly by offering utility fee reductions for managing storm water quality or quantity.[175] Correspondingly, Philadelphia[176] offers green-roof tax incentives of one quarter of the construction cost, thereby removing some market barriers. While city administrations would overlook green-roof zoning, federal agencies, such as the Department of Energy State Energy Program can offer grants for energy efficiency, which green roofs facilitate.

Overall, proponents of green roofs consider them investments for a sustainable future. For example, the American Real Estate Society favors increased worker productivity, reduced sick time, improved employee retention, and overall reduced operating expenses of office buildings with rooftop gardens for energizing and health-promoting breaks.[177] Moreover, green roofs link territorial foresight tools to urban planning by improving local resilience for "future disruptive events that may affect the natural environment and the socioeconomic fabric of a given territory,"[178] such as climate change. Conversely, opponents of green-roof zoning laws may argue that the installation of green roofs is costly. These initial investments average between 3% and 5% more than conventional costs,[179] but

[173] City of Toronto Act, §108 (May 2009) ("The Bylaw requires green roofs on new commercial, institutional and residential development with a minimum Gross Floor Area of 2,000m2 as of January 31, 2010."), http://www1.toronto.ca/wps/portal/contentonly?vgn extoid=83520621f3161410VgnVCM10000071d60f89RCRD&vgnextchannel=3a7a036318061 410VgnVCM10000071d60f89RCRD (last accessed Oct. 19, 2015).

[174] New York City, Mayor's Office of Sustainability, Green Roof Tax Abatement, http://www.nyc.gov/html/gbee/html/incentives/roof.shtml (last accessed Oct. 19, 2015) (noting that "the City of New York and New York State passed legislation in 2008 to provide a one-year tax abatement, or tax relief, of $4.50 per square foot [up to $100,000 or the building's tax liability, whichever is less]. Amended in 2013, the tax abatement is now available through March 15, 2018.").

[175] Minneapolis Public Works, How can you reduce your stormwater fee?, http://www.minneapolismn.gov/publicworks/stormwater/fee/stormwater_fee_stormwater_mngmnt_feecredits (last accessed Oct. 18, 2015).

[176] Phila. Dep't of Revenue, Bus. Privilege Tax—Green Roof Tax Credit, B. 070072 (Apr. 12, 2007), http://beta.phila.gov/media/20160925174047/Green-Roof-Tax-Credit-Overview-Revised-July-20-2016.pdf (last accessed Oct. 18, 2015) (offering "a credit against the Business Privilege Tax of twenty-five percent (25%) of all costs actually incurred to construct the Green Roof, provided that the total credit shall not exceed $100,000.").

[177] Norm Miller et al., Green buildings and productivity, *Journal of Sustainable Real Estate* 1, 1, 65–89 (2009). *See also* Brian W. Edwards and Emanuele Naboni, *Green Buildings Pay: Design, Productivity and Ecology* (Routledge) 2013.

[178] Jose Miguel Fernandez Guell and Leticia Redondo, Linking territorial foresight and urban planning, *Foresight* 14, 4, 316–335, (2012).

[179] Circo, *supra* note 21, at 736.

green roofs pay for themselves through the building's energy savings and longevity.[180] Alternatives to green roofs range from pollution control through cap-and-trade schemes and wait-and-see approaches that maintain the status quo—none of which has satisfactorily mitigated climate change in the past.

Overall, green-roof zoning laws put climate change mitigation control in the hands of communities and provide benefits that extend far beyond their target cities. In following Stuttgart's example, green-roof zoning can mitigate climate change while improving urban resilience to environmental impacts. Future green-roof zoning laws may expand to façade greening to cool buildings vertically. Green-roof zoning offsets urban heat pollution by literally putting "green on top."

Combining these city-greening programs with urban agriculture in new public laws, as outlined in the examples in this chapter, may have promising effects on restoring some agroecology in cities, thereby promoting food integrity.

## 3.4   Summary

This chapter explores how insourcing food production may fix far-reaching problems of the current food system by, for example, deracinating environmentally harmful GMO-dependent monocultures and replacing them with sustainable URFs. The high urban population density provides a consistent customer base and an inherent task force to carry the urban farming models, especially on roofs, which provide "arable land" beyond the real estate competition. Therefore, context-specific holistic upregulation is needed in the sense that existing zoning and local ordinances should not only encourage and incentivize but also reward local food procurement from within the cities to reduce legal barriers and methods of insourcing to feed cities. Through a case study, this chapter explains the connections between the UHIE, climate change and green roofs, introduces the laws underlying Stuttgart's successful city greening, discusses the shortcomings of existing US green-roof policy, and, finally, introduces a model US zoning approach to reduce the UHIE.

[180] Id.

# chapter four

# Agrobiodiversity and agroecology: Contextualizing comparative GMO regulation

This chapter relates agrobiodiversity and agroecology in the context of genetically modified organisms (GMOs) regulation. Selected aspects and consequences of GMO-centric agriculture are compared and explained in terms of their potential to halt or promote food integrity. Through the examples of pollinator protection, invasive species control, and GMO regulatory policy, this chapter brings together various methods of bringing public law to the forefront in the struggle toward achieving food integrity.

Although this chapter zooms out from the metropolitan scope of urban green roof agriculture to mitigate the urban heat island effect and its contributions to global warming, as discussed in Chapter 3, it is complimentary for the legal comparative reasoning of this book. Specifically, by examining how environmental protection is set aside in practices protected by private laws, this chapter highlights the points of attack for public laws. Revisiting Figure 2.1, this chapter focuses on the right side and its considerations of agrobiodiversity.

## 4.1 Pesticide regulation and pollinator protection: Contextualizing how GMOs displace food integrity

Bees make honey, a raw agricultural product that requires comparably little human intervention because it is neither processed nor mixed with other substances before it is sold. Thus, the availability and safety of honey largely depends on bees. But there is more to bees than honey! The buzz of an early spring afternoon, filled with busy bees gathering nectar from the season's first blossoms, cradles people in cozy warmth that summer is nearing. When Rachel Carson, whose landmark book *Silent Spring* spearheaded the environmental movement, paints a picture of still hives, one cannot help but fear a world without bees. Even Albert Einstein warned that "if the bee disappears from the surface of the earth, man would have no more than four years to live."[1] That is true, to some extent, because bees are indicator species, their survival being an index of environmental health and integrity. The greatest threat to their survival and health, in turn, are pesticides, which, widely applied, not only target so-called pests, but also work as biocides for all pollinators, including bees. A careful examination of the chessboard that pesticide producers and users occupy begs the question about the risks of these environmental toxins that are deliberately released into the air. For the reason that GMOs are specifically designed to withstand such toxins, one might change the perspective and note how GMOs exist at the expense of other naturally occurring species. Therefore, bee protection, pesticide regulation, and GMOs are directly linked in and belong in the top right quadrant of Figure 2.1, where GMOs displace agrobiodiversity.

Reconnecting agrobiodiversity, agroecology, and food integrity is at the heart of this multifaceted chapter. All of the following examples have direct effects on agrobiodiversity, thereby on agroecology, and impliedly on food integrity. Another nuance shared in the following examples and analysis is the premise of public versus private law in GMO regulation versus agroecology. Specifically, the shortcomings of regulatory agencies and thus the executive branch illustrate where public law is lacking behind private law, and, thus, how the GMO-centric food industry enjoys privileges that the environment and food system need more urgently in light of increasing climate change–related challenges to food integrity. Considering the chessboard and the perspective shifts introduced in Chapter 1, this chapter uses various relationships, that is, a constellations of facts (the chess figures), to illuminate the problem of trivialization through the proliferation of GMOs and the obstacles to achieving food integrity without improved public law protections.

### 4.1.1   More than honey: EPA and pesticide regulation

A detailed comparison of pesticide regulation in the EU and the United States reveals the perspective clash between the European precautionary and the US biotech approaches. On the metaphorical chessboard, the approaches are so vastly different that they may be depicted as the two colors for players, black and white. In the United States, the black chess figures, the Environmental Protection Agency (EPA) primarily regulates honey production through its pesticide control mandate under the Federal Insecticide, Fungicide, and Rodenticide Act (FIFRA), and secondarily, pursuant to the Toxic Substances Control Act, the Clean Water Act, and the Food Quality Protection Act (FQPA). Specifically, the use of certain

---

[1] Paul Rodgers, Einstein and the Bees. Should You Worry?, *Forbes* (Sept. 9, 2014), https://www.forbes.com/sites/paulrodgers/2014/09/09/einstein-and-the-bees-should-you-worry/#483389098157.

chemicals and substances that present an unreasonable risk of injury to health, wildlife, or the environment[2] are within the EPA's jurisdiction under FIFRA, which empowers EPA to control pesticides.

One of the EPA's tasks under FIFRA is to protect pollinators, such as honeybees, from harmful pesticides.[3] Bees are exposed to harmful pesticides when they forage in fields, forests, parks, gardens, and other green spaces to collect flower nectar, the first step in honey production. Once the bees collect the flower nectar, they break it down into sugars and store it in honeycombs, from which beekeepers later harvest the honey through extraction and centrifugation. When bees forage, they are exposed to systemic pesticides, such as neonicotinoids, which are often used for seed dressings and are particularly dangerous for bees because they are incorporated into the plants, persist and accumulate in soils, and leach into waterways.[4] Moreover, pesticide residues in honey affect public health and endanger food safety.[5] Therefore, the EPA carefully monitors pesticides through its registration process. As part of this process, the EPA follows various steps toward conditional or unconditional pesticide registration, which includes a cost-benefit and scientific analysis to determine whether pesticides are safe for the environment.[6] Only registered pesticides may be used commercially.

Honeybees are dying from these commercially available pesticides. The decline of honeybees in North America has been coined as colony collapse disorder (CCD). Tackling CCD is vital because pollinators, such as bees, are vital for the entire food system, where scientists estimate "that pollinators are responsible for 1 out of every 3 bites of food that we eat."[7] Notably, EPA shares the regulatory space over honey with other agencies to address CCD as part of a CCD Steering Committee, which "consists of scientists from the Department of Agriculture's (USDA) Agricultural Research Service (ARS), National Institute of Food and Agriculture (NIFA), Animal Plant Health Inspection Service (APHIS), Natural Resources Conservation Service (NRCS), Office of Pest Management Policy (OPMP),

[2] Renée Johnson, *The Federal Food Safety System: A Primer*, Congressional Research Service 1–18, 15 (Jan. 17, 2014).
[3] EPA, Protecting Bees and Other Pollinators from Pesticides, http://www2.epa.gov/pollinator-protection.
[4] Dave Goulson, An Overview of the Environmental Risks Posed by Neonicotinoid Insecticides, 50 *Journal of Applied Ecology* 4, 977–987 (Aug. 2013).
[5] Chensheng Lu, Chi-Hsuan Chang, Lin Tao, and Mei Chen, Distributions of Neonicotinoid Insecticides in the Commonwealth of Massachusetts: A Temporal and Spatial Variation Analysis for Pollen and Honey Samples, *Journal of Environmental Chemistry* (online Jul. 23, 2015).
[6] Pollinator Stewardship Council v. U.S. E.P.A., No. 13-72346, 2015 WL 7003600, at *1 (9th Cir. Nov. 12, 2015).
[7] Wayne Buhler, Pollinator Protection, Pesticide Environmental Stewardship, http://pesticidestewardship.org/PollinatorProtection/Pages/default.aspx.

the National Agricultural Statistics Service (NASS)...."[8] This shared regulatory space is another example of what Professor Heinzerling calls opaque and convoluted (see Chapter 3).[9] Finally, the EPA regulates biopesticides, which are types of pesticides derived from such natural materials as animals, plants, bacteria, and certain minerals, such as canola oil and baking soda.[10] Biopesticides are biochemical, microbial, and plant-incorporated protectants (PIPs).[11]

Although the EPA has a special pollinator protection program,[12] many dangerous pesticides are still registered for their prophylactic and broad spectrum use, which, according to Professor Dave Goulson, Britain's foremost bee health expert, "goes against the long-established principles of integrated pest management (IPM), leading to environmental concerns."[13] Therefore, the EPA is largely responsible for the continued survival of honeybees and other pollinators to ensure food safety beyond honey for human enjoyment.

The US regulation of pesticides (in particular, neonicotinoids [neonics]) differs from the EU's treatment, the metaphorical white chess figures on the perspective switchboard. To protect bees from pesticides, the EU has designed one of the strictest approval systems in the world[14] because the European Commission (EC) has recognized the importance of bees to public health, environmental conservation, and economic stability.[15] The EC is the executive body of the EU, headquartered in Brussels, Belgium, and is responsible for proposing legislation, enforcing EU law, and setting

---

[8] USDA, *Report on the National Stakeholders Conference on Honey Bee Health* (Oct. 15–17, 2012), http://www.usda.gov/documents/ReportHoneyBeeHealth.pdf.

[9] Heinzerling, *supra* note xx.

[10] EPA, *Biopesticide Registration*, http://www2.epa.gov/pesticide-registration/biopesticide-registration.

[11] Id.

[12] EPA, *supra* note 3.

[13] Goulson, *supra* note 4.

[14] EC, Pesticides and Bees, http://ec.europa.eu/food/animals/live_animals/bees/pesticides/index_en.htm.

[15] Bees are vitally important for food safety and food security because bees pollinate crops that humans rely on in their diets. Additionally, bees are bio-indicators, meaning that their colony health is indicative of environmental health and, thereby, climate change. Furthermore, bees ensure biodiversity by carrying pollen on; thereby, ensuring genetic diversity in plants and in the organisms relying on those plants for nutrition and habitat. Thus, bees are key players in ecosystem health. Conventional agriculture, especially monocultures, are increasingly dependent on pesticides, synthetic fertilizers, and herbicides that affect bees in a variety of ways, such as by killing them directly, reducing wildflowers on which bees feed, and infringing upon their habitat. Therefore, the EU has initiated some programs to monitor bee health and to assess risks to bee populations in Europe. For more information on the role of bees, please see EFSA's summary with an embedded video, EFSA, Bee Health, http://www.efsa.europa.eu/en/topics/topic/beehealth?wtrl=01.

priorities of action for policy implementation.[16] As such, in 2012, the EC requested that the European Food Safety Authority (EFSA) assess a series of studies on insecticide toxicity to bees, especially on neonics,[17] in response to growing evidence that neonics contribute to bee declines, coined as CCD.[18]

Specifically, the EFSA analyzed three of the most commonly used neonics, "clothianidin, imidacloprid and thiamethoxam as seed treatment or as granules, with particular regard to: their acute and chronic effects on bee colony survival and development; their effects on bee larvae and bee behaviour; and the risks posed by sub-lethal doses of the three substances."[19] According to this analysis, EFSA concluded that neonics were *not* proven safe:

- Exposure from pollen and nectar: Only uses on crops not attractive to honeybees were considered acceptable.
- Exposure from dust: A risk to honeybees was indicated or could not be excluded, with some exceptions, such as use on sugar beet and crops planted in glasshouses, and for the use of some granules.
- Exposure from guttation[20]: The only risk assessment that could be completed was for maize treated with thiamethoxam. In this case, field studies show an acute effect on honeybees exposed to the substance through guttation fluid.[21]

Moreover, the EFSA acknowledges that neonics are toxic to other pollinators, such as birds, bats, and other insects. Notably, neonics pose a "high risk to bees, birds, mammals, aquatic organisms and soil-dwelling organisms," and the available data is insufficient to exclude other dangers associated with neonics.[22] As such, neonics are biocides that directly threaten agrobiodiversity, thereby jeopardizing agroecology and, also, food integrity. Therefore, public law in the EU seeks to protect its territory

[16] European Commission (EC), What We Do, https://www.efsa.europa.eu/en/press/news/120523a.

[17] EC, *supra* note 1.

[18] See *inter alia* Henry et al., A Common Pesticide Decreases Foraging Success and Survival in Honey Bees, 336 *Science* (6079), 348–350 (Apr. 20, 2012); Cresswell and Thompson, Comment on "A Common Pesticide Decreases Foraging Success and Survival in Honey Bees," *Science* 1453 (Sept. 2012); Whitehorn et al., Neonicotinoid Pesticide Reduces Bumble Bee Colony Growth and Queen Production, *Science* 351–352 (April 2012); Reetz et al., Uptake of Neonicotinoid Insecticides by Water-Foraging Honey Bees (Hymenoptera: Apidae) Through Guttation Fluid of Winter Oilseed Rape, *J. Econ. Entomol.* (Oct. 2015).

[19] EFSA, EFSA identifies risks to bees from neonicotinoids (Jan, 16, 2013), http://www.efsa.europa.eu/en/press/news/130116.

[20] EFSA specifies that "[g]uttation is the process by which some plants exude sap in droplets that resemble dew." Id.

[21] Id. (internal citations omitted).

[22] Id.

against neonics—a contrast to the US approach, in which private law protects the interests of neonics producers and users.

Practically speaking, after the EFSA published its findings, the EC restricted the use of neonics by implementing regulation (EU) No. 485/2013. Ultimately, the EC found that "the approved uses of clothianidin, thiamethoxam and imidacloprid no longer satisfy the approval criteria provided for in Article 4 of Regulation (EC) No. 1107/2009 (on plant protection products), with respect to their impact on bees and that the high risk for bees could not be excluded except by imposing further restrictions."[23] Therefore, neonics are not as widely approved in the EU as in the United States.

There are also similarities in some respects to the treatment of GMOs in the United States and the EU. In fact, there are several parallels. For example, the considerations in the EPA's and the EFSA's approvals of neonics and GMOs are similar. It follows that the results of GMOs and neonics proliferation in the United States and the EU correspond to their respective approval by the EPA and EFSA. Additionally, GMOs and neonics are linked, where GMOs in agriculture are genetically engineered crops planted as monocultures and specially designed to have high tolerances for pesticides, such as neonics, thereby creating environmental and economic connections between GMOs and neonics that the EPA and EFSA consider.

On the one hand, as with GMOs, the EPA engages in a cost-benefit analysis, which is in stark contrast to the EFSA's precautionary approach to GMO regulation. In registering pesticides, the EPA considers the economic interests of pesticide producers and use applicants in the approval and registration process under FIFRA. This is similar to the FDA and USDA's biotechnology approach to allow GMOs unless proven unsafe under various food regulation statutes. In contrast, the EU prohibits GMOs unless they are proven safe, pursuant to the precautionary principle.[24] As a result, there is a huge gap in mentality and risk assessment between the EPA's and the EFSA's regulation of neonics, which has led to comparatively broad approval of neonics in the United States but not in the EU. This is, in turn, a contextual and cultural difference that is of vital importance as this comparative analysis continues.

On the other hand, the focus of the EPA's GMO regulation in the United States is split with the FDA, while, in Europe, the EFSA has a more streamlined approach. Here, the EPA regulates pesticides and synthetic

[23] EC Implementing Regulation (EU) No 485/2013 of 24 May 2013 amending Implementing Regulation (EU) No. 540/2011, as regards the conditions of approval of the active substances clothianidin, thiamethoxam, and imidacloprid, and prohibiting the use and sale of seeds treated with plant protection products containing those active substances.

[24] EUR-Lex, Precautonary Principle, http://eur-lex.europa.eu/legal-content/EN/TXT/?uri=URISERV:l32042.

fertilizers, for which GMOs are designed to have high tolerances at the expense of other species growing on monoculture fields, and the FDA focuses on the safety of GMOs as human foods, that is, the result of pesticide-intensive monocultures. Correspondingly, the FDA has little responsibility for bee health and allows food from GMOs to be marketed in the United States. This GMO approval creates a reverse feedback loop that incentivizes pesticide approval for GMO cultivation in agriculture in the EPA's cost-benefit analysis under FIFRA. By comparison, the EFSA is responsible for GMO cultivation and regulation, thereby applying the precautionary approach along the entire way from planting to marketing (or, actually, prohibiting) GMOs.

In sum, the parallels between the EU and US treatment of neonics mimics their respective GMO approval processes. The EPA approves neonics, while the EFSA restricts them. Correspondingly, the FDA, the USDA, and the EPA, within their shared regulatory space, approve GMOs in US agriculture, while the EC generally disfavors GMOs according to the precautionary principle.

Despite all of the similarities, the current debate regarding the impact of US regulation of pesticides (in particular neonics) differs from the EU's treatment. The European debate focuses on "bee health" and the American counterpart on "colony collapse disorder." Thus, while the European approach is bee centered, with bees as an indicator species of environmental health and a key player in food safety, the general public in the United States treats bee decline as a disorder, isolating the problem from the greater picture that includes pesticides and GMOs.

Generally, the European approach is more integrative and comprehensive, albeit focused on (1) wildlife conservation through beekeeping, agriculture, and veterinary surveillance; (2) pesticide use restrictions for public health and food safety; and (3) sustainable agriculture. The EC has even published an infographic that appeals to the general public's empathy for bees and illustrates the great importance of bees for the EU.[25] In fact, Germany even has a specific bee protective law that prohibits substances from endangering bee health.[26]

In the United States, the pollinator protection provision did not even make it into the most recent Farm Bill, and the EPA uses CCD as somewhat of a euphemism for the devastating bee deaths across the country by playing down the underlying problems.[27] Despite science to the contrary, the EPA writes, for example, that,

[25] EC, Bee Health, https://ec.europa.eu/food/animals/live_animals/bees/health_en.
[26] Bienenschutzverordnung (BienSchV), http://www.gesetze-im-internet.de/bienschv_1992/.
[27] EPA, Colony Collapse Disorder, http://www2.epa.gov/pollinator-protection/colony-collapse-disorder.

> [d]ead bees don't necessarily mean CCD. Certain
> pesticides are harmful to bees. That's why we
> require instructions for protecting bees on the
> labels of pesticides that are known to be particu-
> larly harmful to bees. This is one of many reasons
> why everyone must read and follow pesticide label
> instructions. When most or all of the bees in a hive
> are killed by overexposure to a pesticide, we call
> that a beekill incident resulting from acute pesticide
> poisoning. But acute pesticide poisoning of a hive
> is very different from CCD and is almost always
> avoidable.[28]

In conclusion, the debate follows the general legal regulation—or vice
versa. In the EU, consumers are more suspicious of CCD and attempt to
prevent the causes of bee deaths. However, in the United States, a variety
of factors are considered safe unless conclusively proven unsafe for bees.
Although grassroots campaigns try to raise awareness[29] in the United
States, neonics are widely used, and more could be done to protect domestic bees.

Zooming out from bee protection and CCD, bees, as indicator species,
symbolize where the US-American private law protectionism of industry
rights to produce and use pesticides clashes with the EU's precautionary
approach, seeking conservation over economic gain. Table 4.1 illustrates
this comparison.

### 4.1.2    Case study: The new landmark Pollinator
Stewardship Council v. EPA

As described in Chapter 1, this book argues that the proliferation of
GMO-centric agriculture through the trivialization of the associated
risks, a result of the private law protection of the food industry, and the
lack of public law protection for agroecology jeopardize food integrity.
On its flip side, a reduction of GMO-centric agriculture may have the
reverse effect and encourage agroecology. The following case, *Pollinator
Stewardship Council v. U.S. E.P.A.*, exemplifies how this may evolve in the
US common law system. The precedent created herein stifles GMO farming by creating obstacles to the pesticides upon which GMO agriculture
relies. Thinking back to Chapter 2 and the explanation of the risks of

---

[28] Id.
[29] Center for Food Safety, BEE Protective Campaign, http://www.centerforfoodsafety.org/
issues/304/pollinators- and-pesticides/join-the-bee-protective-campaign.

*Table 4.1* Summary of the EU and US approaches to bee health
protection and neonics permitting

|  | EU | United States |
|---|---|---|
| Legislative intent | Bee health and environmental protection | Crop yield stability and pest control |
| Agency responses | EFSA's prohibition of neonics | EPA's permission under cost-benefit analysis |
| Rationale for legislation | Integrative and comprehensive risk assessment | Industry-reliance and limited risk examination |
| Factual construction | Neonics are not proven safe and therefore prohibited | Neonics to be used despite environmental safety concerns |

GMO-focused monocultures reminds the reader that any point of attack may have a positive ripple effect on the path toward food integrity.

Several non-profit and watchdog organizations have touted pesticides as the foremost threats to pollinators, but case law is only slowly trickling in. Now that some landmark cases have been decided by US courts, they create precedent that involves the jurisdiction as the last branch of government to follow suit in the issue of pesticide toxicity for the environment and, thereby, food supplies. In *Pollinator Stewardship Council v. U.S. E.P.A.*, the Court of Appeals for the Ninth Circuit vacated the EPA's unconditional registration of sulfoxaflor.

Sulfoxaflor is an insecticide that acts on the same receptor in insects as the class of insecticides referred to as neonicotinoids (neonics) does, which are highly toxic to bees. Neonics, including sulfoxaflor, are "systemic" insecticides, which cause insects to die when they come into contact with the pesticide, sprayed onto or absorbed by the plant. Respondent-Intervenor Dow asked the EPA to approve sulfoxaflor for use on a variety of different crops, including citrus, cotton, fruiting vegetables, canola, strawberries, soybeans, wheat, and many others. The petitioners, Earthjustice, appealed the EPA's decision to register sulfoxaflor unconditionally. Amici Curae[30] included several distinguished environmental advocacy groups, which voiced their concerns about the EPA's approval of sulfoxaflor.

The issue on appeal was whether the EPA's unconditional registration under FIFRA was supported by the record as a whole, where the studies

---

[30] Amici Curiae included: Center for Food Safety, Northeast Organic Farming Association Interstate Council, Northeast Organic Farming Association, Massachusetts Chapter, Inc., Northeast Organic Farming Association of Rhode Island, Inc., Northeast Organic Farming Association of New York, Inc., Maine Organic Farmers and Gardeners Association; Defenders of Wildlife, Friends of the Earth, Center for Environmental Health, Conservation Law Foundation, Midwest Organic and Sustainable Education Service, Beyond Pesticides, Pesticide Action Network of North America, The Sierra Club, National Family Farm Coalition, and American Bird Conservancy.

underlying the EPA's Tier 2 analyses had serious limitations. FIFRA pro-hibits the use or sale of pesticides that lack approval and registration by the EPA,[31] which may deny an application for registration when "necessary to prevent unreasonable adverse effects on the environment."[32] Specifically, FIFRA prohibits growers from using or companies from selling any pes-ticide that the EPA has not approved and registered.[33] Thus, the EPA con-ditionally registers a pesticide[34] when there is insufficient data to evaluate the environmental effects of a new pesticide, permitting the pesticide to be used "for a period reasonably sufficient for the generation and submis-sion of required data."[35] Unconditional registration, however, necessar-ily requires sufficient data to evaluate the environmental risks. Here, the court held that the unconditional registration was unsupported because the EPA's decision was based on flawed and limited data. Therefore, the EPA's unconditional registration of sulfoxaflor was vacated and remanded for the EPA to obtain further studies and data regarding the effects of sulfoxaflor on bees.

Shifting the perspective away from this one undoubtedly significant case to the larger picture it may serve, one must acknowledge how far the reach of pollinators truly is. As Larissa Walker and Sylvia Wu, experts from the Center for Food Safety, explain,

> [p]ollinating insects, such as bees, butterflies, birds, bats, and other animals are critical to maintaining healthy ecosystems and a strong agricultural econ-omy. These important invertebrates ensure repro-duction, fruit set development and seed dispersal in the vast majority of plants, both in agricultural landscapes and natural ecosystems. Pollinators con-tribute an estimated $20–30 billion annually to the U.S. agricultural economy. Bees and other pollina-tors also support the reproduction of nearly 85% of the world's flowering plants—more than 250,000 varieties globally.[36]

---

[31] 7 U.S.C. § 136a(a).

[32] Id.

[33] Id.; FIFRA uses a "cost-benefit analysis to ensure that there is no unreasonable risk cre-ated for people or the environment from a pesticide." Washington Toxics Coal. v. EPA, 413 F.3d 1024, 1032 (9th Cir.2005).

[34] FIFRA allows EPA to deny an application for registration of a pesticide to prevent "unrea-sonable adverse effects." EPA may either "unconditionally" register a pesticide, 7 U.S.C. § 136a(c)(5), or "conditionally" register it, § 136a(c)(7)(C).

[35] Id.

[36] Larissa Walker and Sylvia Wu, Pollinators and pesticides in *International Farm Animal, Wildlife and Food Safety Law 496* (Gabriela Steier and Kiran Patel, Eds.). Springer (2017) (internal citations omitted).

Consequently, bees and other pollinators are cornerstones of a functioning food system. Placing them anywhere in the right side of Figure 2.1, they are somewhat of a lubricant to make the cycle go around, while the perpetuating cycle threatens their existence.

### 4.1.3 Enlightened utilitarianism for cooperating species: Comparison to seek solutions

Protecting bees to ensure food safety might be one of the most powerful methods to stop the proliferation of GMOs, with all of its devastating ripple effect on food safety and the environment. Humans and bees are cooperating species, where humans provide foraging opportunities on agricultural fields for bees, which, in turn, pollinate crops. In fact, the Food and Agriculture Organisation (FAO) of the United Nations "estimates that out of some 100 crop species which provide 90% of food worldwide, 71 of these are bee-pollinated."[37] Bees are indicator species and signal environmental health, which is necessary to ensure food safety,[38] as noted earlier in this chapter. Agricultural pesticide uses, however, cause alarming bee declines and correlate with environmental degradation. Conversely, German and Chinese pesticide regulations diverge in protecting the coexistence of humans and bees for food safety.

Germany has an extensive plant protective law, the *Pflanzenschutzgesetz*,[39] which enables a special bee protective law through §3 Article 1, the *Bienenschutzverordnung* (BienSchV). This federal law protects bees from pesticides to ensure food safety. BienSchV outlines extensive testing requirements of bee toxicity prior to pesticide registration, labeling laws with varying degrees of bee toxicity, pesticide use restrictions, and penalties for violations. Dating back to 1938,[40] the BienSchV showcases Germany's deep commitment to bee conservation.

In contrast, Chinese law offers no counterpart to the German BienSchV. Despite the Chinese government's campaign against banned pesticides that are still produced and used,[41] no specific statute protects bees. Only

[37] The United Nations Environment Programme (UNEP), Global Honey Bee Colony Disorder and Other Threats to Insect Pollinators (2010), https://wedocs.unep.org/bitstream/handle/20.500.11822/8544/-UNEP%20emerging%20issues_%20global%20honey%20bee%20colony%20disorder%20and%20other%20threats%20to%20insect%20pollinators-2010Global_Bee_Colony_Disorder_and_Threats_insect_pollinators.pdf?sequence=3&isAllowed=y (internal citations omitted).

[38] Wu et al., Research progress in bee and honeybee product as biological indicator for monitoring environmental pollution, *J. Agric. Sci. Tech.* 3 (2008), http://en.cnki.com.cn/Article_en/CJFDTOTAL-NKDB200803003.htm.

[39] *Pflanzenschutzgesetz (PflSchG)*, http://www.gesetze-im-internet.de/bundesrecht/pflschg_2012/gesamt.pdf (translation by author).

[40] Julius-Kühn Institut, *Bienenschutzverordnung*, https://www.gesetze-im-internet.de/bienschv_1992/BienSchV_1992.pdf (translation by author).

[41] Beyond Pesticides, *China Works to Improve Food Safety Image* (Aug. 9, 2007), http://www.beyondpesticides.org/dailynewsblog/2007/08/china-works-to-improve-food-safety-image/.

vague progress emerges where "[t]he latest draft amendment to the Food Safety Law, submitted to the National People's Congress for approval, encourages development of less toxic pesticides to replace highly toxic ones and forbids the[ir] use ... in fruit and vegetable farms...."[42] This Chinese legislative gap jeopardizes food safety.

These observations are dependent on cultural differences, as well as on differing economic goals, although it has been established that pesticide and fertilizer use in modern farming threaten bees' survival and, thereby, the agricultural ecosystem services they provide.[43] Professor Dave Goulson, a leading expert on bees, documents bee declines and extinction in Europe and China and mainly blames neonicotinoid pesticides while predicting that agricultural yields correlate with bee declines.[44] In fact, China's uncontrolled pesticide uses are evident in "the apple and pear orchards of south west China, where wild bees have been eradicated by excessive pesticide use"[45] and where "farmers have been forced to hand-pollinate their trees...."[46] However, An Jiandong, a researcher from the Department of Apiculture at the Chinese Academy of Agricultural Sciences, notes that no specific data exists to determine the true losses.[47] Goulson also emphasizes that "there are not enough humans in the world to pollinate all of our crops by hand"[48] and, therefore, bees need protection.

As compared to China, and thanks to BienSchV, Germany more successfully protects bees.[49] German public interest law, by way of the BienSchV, implements EU-wide legislation,[50] which accomplishes some enlightened utilitarianism.[51] Notably, the EC "[c]onsiders it important to take urgent measures to protect bee health..."[52] and banned many bee-toxic pesticides.[53] In contrast, the Chinese parliament takes the opposite approach and permits blacklisted pesticide uses in large amounts. In fact, Yang Zhen, a member of the NPC Standing Committee, stated outright

[42] China Daily, Phasing out the use of toxic pesticides, *China Daily* (Apr. 22, 2015), http://www.chinadaily.com.cn/opinion/2015-04/22/content_20502194.htm.

[43] UNEP, *supra* note 7.

[44] Dave Goulson, Decline of bees forces China's apple farmers to pollinate by hand, *China Dialogue* (Feb. 10, 2012), https://www.chinadialogue.net/article/show/single/en/5193-Decline-of-bees-forces-China-s-apple-farmers-to-pollinate-by-hand.

[45] Id.

[46] Id.

[47] Harold Thibault, When humans are forced to replace the bees they killed, *Worldcrunch* (May 5, 2014).

[48] Id.

[49] UNEP, *supra* note 7.

[50] Directive 2009/128/EC (21 October 2009).

[51] Stanford Encyclopedia of Philosophy, Herbert Spencer (Sep 17, 2012), http://plato.stanford.edu/entries/spencer/.

[52] Resolution (2011/2108(INI)).

[53] Li Yang, Ban on highly toxic pesticides should be the first step, *China Daily* (Apr. 23, 2015), http://www.chinadaily.com.cn/opinion/2015-04/23/content_20515155.htm.

that "it is not possible to totally ban the use of highly toxic pesticides."[54] Thus, the Chinese neglect environmental conservation to promote economic growth—a near-sighted goal from a German point of view.

In conclusion, the German population equilibrium approach better ensures the survival of cooperating species. The Chinese growth mentality[55] appears outdated in the bee protective context because it undermines food safety for humans. Rather than compete, populations such as humans and bees can coexist to adapt to environmental changes, thereby persisting in the face of hunger, environmental degradation, and extinction.[56] Long-term initiatives to ensure food integrity may better protect economic and environmental interests, combining the objectives of both the German and the Chinese cultural contexts for their respective bee protective laws (or the lack thereof).

## 4.2 Alternative biodiversity management: Controlling invasive species

Whereas bees in the previous section are firmly rooted in managed agriculture, providing crucial pollination services to ensure that crops of flowering plants thrive, other species are considered pests. By definition, a pest is an unwanted and troublesome creature, often an insect that harms cultivated plants, or a rodent that annoys humans. However, pesticides attempt to tackle these pests and have been failing for centuries because rats, ants, mice, cockroaches, lice, and other similarly unwelcome creatures thrive around the globe. The GMO-centric monocultures, nonetheless, follow the outdated and disproven idea that poisoning pests increased crop yields. Culturally, the United States responds with more and heavier artillery, highly toxic pesticides, while the EU seeks agroecologic solutions, to some extent. Thus, the following section proposes another perspective shift, suggesting that a change in attitude, from considering that pests are useless to appreciating their usefulness and manageability, could be a proactive way to tackle the much-lamented problem of their numerous existence. So, why not eat pests?

On the flip side of the following thought-experiment lies support for the proposition that eating alleged pests could also help to stop the problem of pesticide use and the proliferation of specially designed high-pesticide and herbicide tolerant crops, that is, GMOs, whose risks to the environment are trivialized. Simply put, if one were to eat the species that

---

[54] Yang Zhen, cited in *China Daily, supra* note 6.
[55] Wang Canfa, Chinese Environmental Law Enforcement: Current Deficiencies and Suggested Reforms, 9 *VT. J. Env. L.* 161–173 (2007).
[56] Greg Graffin, *Population Wars: A New Perspective on Competition and Coexistence*, at 13. St. Martin's Press (2015).

supposedly threaten harvests, one might eliminate the need for pesticides, thereby eliminating the usefulness of GMOs, which may weaken their patentability, marketability, and overall market. The results may be quite promising, support agroecology, and help to achieve food integrity in a multitude of ways.

### 4.2.1   One person's pest is another's meal[57]: Tweaking food safety regulation to manage agrobiodiversity

Tweaking food safety regulation to manage agrobiodiversity provides a viable avenue to protect native biodiversity even though, initially, the idea seems preposterous to the Western palate: Eating invasive species.[58] On second thoughts, however, it turns out to be a sustainable strategy and culinary adventure that might help to restore ecologic balance—and it could be quite tasty. An invasivore's menu may begin with a mitten crab cake or zebra mussel plate on freshwater algae, continue with a hearty feral hog loin with field mustard crudités or rabbit ragout garnished with dandelion pesto, and end in a blackberry sorbet. This menu, however, is not properly regulated by any of the 13 federal agencies overseeing invasive species and food safety, or any of the 6 applicable federal statutes (see Chapter 2).[59] Ensuring the safety of such a menu is, nonetheless, an important prerequisite to endorsing invasivorism as a pest control mechanism to protect agrobiodiversity.

Regulating invasive species in ways other than through environmentally harmful pesticides is an issue ripe for legal review because agrobiodiversity is more threatened than it has ever been before.[60] However, it may not seem to fit within the overarching federal regulatory system. Domesticated plants and animals cannot compete with invasive species, which threaten the ecosystems on which agriculture depends.[61] Thus, "eat the invaders—or let them eat your food,"[62] becomes a pest control strategy,

---

[57] Title adapted from Gregory Han, One Man's Pest, Another Man's Meal: Where to Grub on Bugs in Los Angeles, KCET Food (Aug. 11, 2014), http://www.kcet.org/living/food/the-nosh/where-to-eat-bugs-in-los-angeles.html.

[58] Executive Order 13112 (EO 13112) (Feb. 3, 1999), http://www.invasivespeciesinfo.gov/laws/execorder.shtml, defines an invasive species as, "an alien species whose introduction does or is likely to cause economic or environmental harm or harm to human health."

[59] National Environmental Policy Act of 1969 (42 U.S.C. 4321 *et seq.*), Nonindigenous Aquatic Nuisance Prevention and Control Act of 1990, (16 U.S.C. 4701 *et seq.*), Lacey Act, (18 U.S.C. 42), Federal Plant Pest Act (7 U.S.C. 150aa *et seq.*), Federal Noxious Weed Act of 1974, (7 U.S.C. 2801 *et seq.*), Endangered Species Act of 1973, (16 U.S.C. 1531 *et seq.*)

[60] Mark A. Davis, Biotic globalization: Does competition from introduced species threaten biodiversity?, *BioScience* 53 (5) 481–489 (2003).

[61] Joshua Galperin and Sara Kuebbing, Eating Invaders: Managing Biological Invasions with Fork and Knife?, *Nat. Resources & Env't* 41 (2013).

[62] Id.

not just a culinary adventure. Therefore, it also poses a food safety concern, specifically targeting local agriculture and food production that is beyond the federal regulatory purview.

Locally targeted state agency action should address invasivorism because it could be an environmentally friendly and nutritious pest control strategy if its food safety were ensured. Then Section 4.2.2 discusses the federal food safety regulatory shortcomings of invasivorism and Section 4.2.3 proposes regulatory improvements. Section 4.2.4 discusses the food safety obstacles of regulating invasive species. Ultimately, the chapter concludes with future considerations for invasivorism as a pest management strategy (IPMS) to protect agrobiodiversity in global trade. Combined, the examples and analysis in this chapter illustrate how public law can achieve what private law has not been able to—the promotion of food integrity.

## 4.2.2   Inadequate regulation of agricultural invasive species in the US food safety system

Edible invasive species should be locally regulated as a matter of food safety because they would offer more sustainable and, arguably, safer pest control measures. The current regulation on invasive species fails to address invasivorism as a pest control and food safety concern because, on the one hand, current federal invasive species regulation is fragmented, convoluted, and misguided. On the other hand, none of the current regulatory schemes sufficiently address the food safety challenges of invasive species. These food safety perspectives combine foodborne illness, agricultural sustainability, and environmental concerns, but are neglected by the regulatory scheme now in place—an example of the shortcoming of public law to-date.

Currently, the National Invasive Species Council (NISC), established through Executive Order 13112,[63] distributes federal invasive species management regulation among 13 agencies. Each agency's secretary co-chairs the council. The council, in turn, oversees the NISC staff and collaborates with the Invasive Species Advisory Committee. The NISC's duties include international, national, state, and local invasive species management. As one of these goals, the NISC issues a national Invasive Species Management Plan (ISMP) every five years.

---

[63] Executive Order 13112, *supra* note 2.

### 4.2.3 Locavorism: Improving regulation of agricultural invasive species in the US food safety system

The regulation of agricultural invasive species in the US food safety system would benefit from local invasive species tracking and educational outreach. Local invasive species tracking would be in line with the second and third goals under the current ISMP: early detection, rapid response, control, and management.[64] Similar measures have been successful in Florida's Fish and Wildlife Conservation Commission's new app that tracks the gopher tortoise, a threatened species, and Ohio State University's app for reporting sightings of invaders in the Great Lakes, like Asian carps and long-horned beetles.[65] Using such online tracking systems through websites or cell phone applications, for example, may allow the public to find and report edible invasive species. Consequently, local regulatory authorities might dispatch field officers to take samples, submit them to testing, and publish warnings if any of those species are unsafe for human consumption.

Alternatively, local food safety and wildlife management agencies' educational outreach could publish lists of edible invasive species, offer schooling programs to educate consumers how to find, harvest, or hunt them, and how to safely handle and prepare them for human consumption. Backed by the resources of the federal agencies, local governments could target resources directly to consumers, to ensure the greatest possible impact and highest food safety. In fact, in 2010, the National Oceanic and Atmospheric Administration (NOAA) launched such an educational campaign in Jamaica to interest consumers in eating lionfish. James Morris, an ecologist with NOAA, said that "[l]ionfish are really great eating."[66] The 66% drop in lionfish sightings[67] proves the campaign's success in the lionfish restaurant market.[68] Educational outreach could also encompass guidelines for the safe processing of wild game[69] or the handling of wildflower honey. Guidelines for safely processing feral hogs[70] could serve as viable examples and inform educational outreach initiatives.

---

[64] 2008–2012 Invasive Species Management Plan (Aug. 1, 2008), 1–35, 16, 21, http://www.invasivespeciesinfo.gov/council/mp2008.pdf.

[65] Casey N. Cep, An App to Find Nemo, *The New Yorker* (Jul. 22, 2014), http://www.newyorker.com/tech/elements/app-hunts-fish-2.

[66] Id.

[67] Id.

[68] Id.

[69] See, for example, Ohio Rev. Code Ann. § 1531.101 (West) ("Acceptable methods of taking migratory game birds"); N.C. Gen. Stat. Ann. § 113-291.1 ("Manner of taking wild animals and wild birds"); Ga. Code Ann. § 27-1-3 (West) ("hunting and fishing a wildlife regulations").

[70] Michelle Greenhalgh, Guidelines for Safely Processing Wild Game, *Food Safety News* (Nov. 2, 2010), http://www.foodsafetynews.com/2010/11/michigan-hunters-reminded-of-safe-game-handling-practices/#.VlI5o7QeO0Y.

## 4.2.4 Food safety challenges of proportional regulation of agricultural invasive species

The challenges to proportional regulation of invasive species from the food safety perspective are (1) farming popular invasive species for food may spread the pests,[71] (2) cultural obstacles to eating pests, (3) food contamination of unregulated wild game, and (4) feasibility. First, if invasivorism did catch on, some people might begin to cultivate invasive species instead of the intermediate step of hunting and gathering them. This would undermine the purpose of invasivorism as pest control. Consequently, to stop invasive species cultivation for financial gain, local food safety authorities may prohibit the farming of cultivated invasive species, such as Asian carp or lionfish, and monitor the sales or processing of invasive species carefully. Such localized agency regulation of invasivorism would also benefit from the tracking and educational outreach programs because it would inform the public about the important pest control purpose of invasivorism. This set of procedures would be a call for heightened public law support of invasivorism.

Second, depending on cultural differences and dietary preferences, some edible invasive species may sound less appetizing, albeit edible, such as rabbit, bullfrog, nutria (river rat), and lionfish.[72] Therefore, invasivorism may not effectively function as a pest control strategy if people refuse to eat those species. It follows that only some species, such as carp, catfish, fennel, crabs, wakame seaweed, and watercress may be targeted through invasivorism. Although this might seem to shortchange the proposed solution, the opposite is true: Targeting any invasive species through local consumer-driven action will have a positive impact on pesticide and herbicide reduction.[73]

Third, food safety concerns of invasivorism touch upon the wildlife–livestock interface and heavy metal or chemical pesticide, and antibiotic contamination of meat. Wild animals, such as game birds, may carry a multitude of zoonoses that may also infect humans. For instance, the direct transmission of zoonoses from domesticated animals to wildlife and to humans, or vice versa, at the wildlife–livestock interface poses food safety threats. Where "[n]early 80% of the pathogens present in the United States have a potential wildlife component," and human "encroachment into wildlife habitat continues to increase along with more intensified livestock production practices," veterinary scientists warn that the "potential for contact and pathogen transmission at the

---

[71] Science Gallery, Invasivorism, https://dublin.sciencegallery.com/edible/place/invasivorism/.
[72] Eat The Invaders, http://eattheinvaders.org.
[73] The unpalatable invasive species are beyond the scope of this chapter and may be targeted through other measures, such as introducing natural predators.

livestock–wildlife interface" is increasing.[74] Moreover, invasive species, such as feral hogs and deer, feed on various uncontrolled sources, and an invasivore's diet may, therefore, be high in heavy metals, ingested with the meat of these animals.

In the grand scheme of things, this is, nonetheless, the lesser of two evils. Conventional diets consisting of staple farmed foods, such a pesticide-dipped berries, antibiotic-spiked pork, or, nowadays, GE salmon,[75] are no healthier options. In fact, a 2010 Food Safety and Inspection Service (FSIS) Audit Report revealed that domesticated beef may also exceed safety thresholds and concluded that the USDA-FSIS-administered national residue program unsuccessfully monitors these food safety threats.[76] As David Kirby, investigative journalist, highlights, "some US beef is contaminated with heavy metals like copper and arsenic, antibiotics like Flunixin, penicillin, and Ivermectin, and a host of pesticides—all of which are used in concentrated animal feeding operations (CAFOs)."[77] The public health and environmental repercussions of consuming and, thereby, supporting the production and sale of factory farmed meat[78] have devastating consequences that might supersede those of wild game. Moreover, the consumption of meat from CAFOs is a direct trigger of the proliferation of GMOs, which are used as feed and fodder, or for processed foods, which, again, trivialize their dangers by shrouding them in secrecy under private law protection.

Finally, the consumer friendliness of hunting or gathering invasive species is another obstacle. As Casey Cep, writing for the *New Yorker*, notes, "even when there's a taste for these non-native plants and animals, very few of them can be commercially harvested. Individuals often have to hunt or gather these critters on their own, which is especially difficult to do with lionfish, since there are no effective traps or bait."[79] Nonetheless, various solutions have been successfully field tested. For example, in 2012, Florida "waived fishing-permit requirements for divers

[74] Miller et al., Diseases at the Livestock–Wildlife Interface: Status, Challenges, and Opportunities in the United States, *Preventive Veterinary Medicine* 110(2), 119–132, http://www.sciencedirect.com/science/article/pii/S0167587712003984.

[75] FDA, AquAdvantage Salmon, http://www.fda.gov/AnimalVeterinary/DevelopmentApprovalProcess/GeneticEngineering/GeneticallyEngineeredAnimals/ucm280853.htm.

[76] USDA-FSIS National Residue Program for Cattle, Audit Report 24601-08-KC (Mar. 2010), http://www.usda.gov/oig/webdocs/24601-08-KC.pdf.

[77] David Kirby, Drugs, Poisons and Metals in Our Meat—USDA Needs a Major Overhaul, *Huffington Post* (Jun. 14, 2010), http://www.huffingtonpost.com/david-kirby/food-safety----drugs-pois_b_537686.html.

[78] Id. See also Doug Gurian-Sherman, CAFOs Uncovered, Union of Concerned Scientists (Apr. 2008), http://www.ucsusa.org/sites/default/files/legacy/assets/documents/food_and_agriculture/cafos-uncovered.pdf.

[79] CEP, *supra* note 15.

who go after lionfish, and it removed the bag limit on the species."[80] Half a million dollars in funding are pending governor approval.[81] Alternatively, Cep proposes invasive species-specific bounties, derbies, and tournaments, which "have removed thousands of fish in only a few hours."[82] Notably, "the University of Southern Mississippi offered a five-dollar reward for every lionfish caught in the Gulf of Mexico west of Mobile Bay, and derbies sponsored by the Reef Environmental Education Foundation … have removed more than twelve thousand lionfish since 2009."[83] Consequently, the inconvenience obstacles of invasivorism can be overcome through incentives.

In conclusion, invasivorism may be a proactive method of decimating pests without resorting to chemical and synthetic pesticides, and it would even add to a varied, locally sourced diet, protect agrobiodiversity, and potentially be entertaining. However, there are several obstacles in linking pest management to food safety within invasivorism. The public currently bears the cost of over $137 billion for corrective measures to combat invasive species.[84] These costs are likely on the rise with increasingly centralized food production and evolving global trade of foods, which facilitate invasive species hitchhikers.[85] Nonetheless, IPMS could protect agrobiodiversity despite global trade, if locally targeted state agency action better regulated invasivorism food safety.

## 4.3   Summary

This chapter relates agrobiodiversity and agroecology in the context of GMO regulation. Selected aspects and consequences of GMO-centric agriculture are compared and explained in terms of their potential to halt or promote food integrity. Through the examples of pollinator protection, invasive species control, and GMO regulatory policy, this chapter brings together various methods of bringing public law to the forefront in the struggle toward achieving food integrity. This chapter begins with the federal food safety regulatory shortcomings of invasivorism and continues with proposed regulatory improvements. Then, it discusses food safety obstacles of regulating invasive species. Ultimately, the chapter concludes with future considerations for invasivorism as an IPMS to protect

---

[80] Id.
[81] Id.
[82] Id.
[83] Id.
[84] James S. Neal McCubbins et al., Frayed Seams in the "Patchwork Quilt" of American Federalism: An Empirical Analysis of Invasive Plant Species Regulation, 43 *Envtl. L.* 35, 81 n.4 (2013) (internal citations omitted).
[85] 2008–2012 National Invasive Species Management Plan at 3.

agrobiodiversity in global trade. The chapter concludes that invasivorism may be a proactive method of decimating pests without resorting to chemical and synthetic pesticides, and it would even add to a varied, locally sourced diet, protect agrobiodiversity, and be potentially entertaining. However, there are several obstacles in linking pest management to food safety within invasivorism.

*chapter five*

---

# The regulation of GMOs
# in international trade

## 5.1  Cropping GMO regulation: US regulation
## as compared to EU regulation

One may ask why nobody stops this self-undermining system that imperils public health and environmental integrity, and why nobody stands up against the agricultural industry (BigAg) and for food integrity? How the system centers around BigAg, which is illustrated in Figure 2.1, is the subject of Section 5.1.1. Using the case study of Agent Orange Corn, Section 5.1.2 shows how genetically modified organisms (GMOs) are integrated into the food regulatory network in the United States. In response to all those who wonder why nothing is being done to stop the proliferation of GMOs, a brief answer would be that one needs information upon which to act, and the industry does not easily relinquish control of this information. Uncovering the dirty business of industrial agriculture would ruin people's appetites, thereby diminishing BigAg's profits. Shifting one's perspective from the industry's release of information, one may reach the point of the public's demand for information—the right to know. This right to know embodies a forced release of information, which in food law is simply called labeling and the subject of Section 5.1.3. The second part of this chapter zooms in on transgenic animals, the first GMOs that are approved in the animal kingdom: patented fish. Developing the power of the right to know and product labeling further, Sections 5.2.1 through 5.2.4 provide information about the consumer-oriented controls to create points of attack on the proliferation of GMOs and the corresponding trivialization of the risks of a GMO-dominated food system (see Figure 2.1).

## 5.1.1   Neoliberal farm bill idea: New frontiers in sustainability

An oxymoron based in capitalism: The current neoliberal[1] mega-agribusiness (BigAg) in the United States endangers the environment and fosters climate change—it undermines itself by destroying the natural resources upon which agriculture depends. Sustainable development, however, provides a competitive advantage to the agribusiness sector and is a necessary alternative to the neoliberal agricultural growth under the Farm Bill with the potential to reverse environmental harm. Therefore, if BigAg foils sustainable growth in the agricultural practices under the current Farm Bill, the food producing sector may develop a competitive advantage and foster innovation and environmental integrity. At the same time, this sustainable growth would benefit economic growth in a climate-friendly democracy.

Shrouded by the Farm Bill, agrarian democracy clashes with neoliberal creative destruction of the environment and the economy, thereby jeopardizing the future of agriculture as a whole.

As the distinguished environmental lawyer, William Eubanks, writes, "[t]he 1933 Farm Bill was designed to save small farming in America, and it signaled a return to the Jeffersonian ideal of an agrarian democracy."[2] In reality, however, the Farm Bill has fostered a culture of creative economic and environmental destruction[3] under the Green Revolution[4] by

[1] The dictionary definition of neoliberal is "relating to or denoting a modified form of liberalism tending to favor free-market capitalism." Noam Chomsky, a professor at MIT, commented on neoliberal capitalist rule in *Profit Over People: Neoliberalism and Global Order*. In the introduction to this landmark work, Robert W. McChesney, professor at the University of Illinois at Urbana–Champaign and co-founder of Free Press, explained that "Neoliberalism is the defining political economic paradigm of our time—it refers to the policies and processes whereby a relative handful of private interests are permitted to control as much as possible of social life in order to maximize their personal profit." Robert W. McChesney, in Noam Chomsky, *Profit Over People: Neoliberalism and Global Order* (2011), Kindle location 27.

[2] William Eubanks, Paying the farm bill. *Environmental Forum* 27, 56–75, 58 (2010).

[3] Creative destruction is an economic term, "coined by Joseph Schumpeter in his work entitled 'Capitalism, Socialism and Democracy' (1942) to denote a 'process of industrial mutation that incessantly revolutionizes the economic structure from within, incessantly destroying the old one, incessantly creating a new one.'" Creative Destruction definition, *Investopedia*, http://www.investopedia.com/terms/c/creativedestruction.asp#ixzz3oCXY7eWb.

[4] Eubanks explains that "the Green Revolution led to plant breeding and hybridization and military technology developed during World War II led to new pesticides, herbicides, and agricultural mechanization. These advances increased yields, which resulted in overproduction and depressed crop prices." Eubanks, *supra* note 2. He also links the Green Revolution to the agriculture lobby that greatly influenced the Farm Bill, where "larger farms that had the ability to stay afloat despite decreased crop prices began to exploit the weaker, smaller outfits by purchasing foreclosed operations at below-market rates and by joining forces with other large farms and food processors to create the agribusiness lobby." Id. at 58 (internal citations omitted).

killing agricultural integrity from within and replacing it with toxic farming practices. These toxic farming practices include, for example, toxic water discharges in violation of the Clean Water Act.[5] For the environment, the resulting creative destruction coincides with replacing small, family-owned farms with large agribusiness operations engaging in toxic farming practices.[6] Correspondingly, the economic destruction lies in the subsidization of commodity crop price dumping and the failure to use taxes to preserve national environmental integrity. Thus, the agribusiness sector, in implementing and shaping the current Farm Bill, undermines the economic system upon which its neoliberal drive for capital growth rests by jeopardizing the environment. As a result, the Farm Bill artificially distorts the market,[7] while threatening the environment through unsustainable farming and weakening its chances for continued growth. In other words, the Farm Bill exemplifies the private versus public law clash previously noted, while embodying a seemingly public piece of legislation with private law underpinnings that protect the industry rather than the public and the environment. The Farm Bill is somewhat of an antithesis to public laws that may promote food integrity, an exemplary lesson in what not to do.

A reversal of the market distortions originating from Farm Bill policy may be possible if sustainability played a greater role under the Farm Bill. The obstacle to this reversal lies in corporate error and disruptive innovation, where "most executives treat the need to become sustainable as a corporate social responsibility, divorced from business objectives."[8] In the neoliberal American agricultural sector, "the fight to save the planet has turned into a pitched battle between governments and companies, between companies and consumer activists, and sometimes between consumer activists and governments."[9] As a product of this battle, the Farm Bill has, on the one hand, facilitated an oxymoronic practice of using tax dollars to subsidize commodity crop price dumping, thereby impoverishing small farmers and promoting unsustainable farming.[10] On the other hand, the Farm Bill's apparent efforts to conserve the environment have

---

[5] Waterkeeper Alliance, Inc. v. Hudson, No. CIV.A. WMN-10-487, 2012 WL 6651930, at 16 (D. Md. Dec. 20, 2012).

[6] Id.

[7] Eubanks, *supra* note 2, at 58.

[8] Ram Nidumolu, C.K. Prahalad, and M.R. Rangaswami, Disruptive innovation: Why sustainability is now the key driver of innovation, *Harvard Business Review* (Sept. 2009), https://hbr.org/2009/09/why-sustainability-is-now-the-key-driver-of-innovation.

[9] Id.

[10] Eubanks, *supra* note 2, at 60 (explaining that "just five crops—corn, cotton, rice, soybeans, and wheat—control the commodity subsidy market . . . Despite the fact that 'thousands of plant and animal species are cultivated for human use,' more than 84 percent of the $172 billion spent to subsidize our nation's agriculture during that period went solely to these five crops.").

"failed because of a lack of funding, conflicts of interest, a poor conservation payment structure, and a failure of environmental laws to include enforcement mechanisms against the agriculture industry for violation of those laws."[11] Thus, agribusiness executives must step up and foster sustainable agricultural innovation to offset the negative consequences of the current Farm Bill. Corporate goals to boost the bottom line, however, can promote innovation for sustainable growth as an economic strategy that will also benefit the environment and ensure future food security by reversing toxic farming practices.

In fact, environmental conservation is the cornerstone of sustainable growth. For instance, a recent study found,

> that sustainability is a mother lode of organizational and technological innovations that yield both bottom-line and top-line returns. Becoming environment-friendly lowers costs because companies end up reducing the inputs they use. In addition, the process generates additional revenues from better products or enables companies to create new businesses. In fact, because those are the goals of corporate innovation ... smart companies now treat sustainability as innovation's new frontier.

Thus, according to these findings, the mega business agricultural giants that took over the American small farms would also benefit from prioritizing sustainability. Sustainability will equip forward-thinking agribusiness companies with a competitive advantage over their rivals.[12] From a different perspective, switching the metaphorical chessboard, focusing private laws on sustainability may unite the goals of both private and public legislation.

In conclusion, sustainable development in the agribusiness sector will not only foster environmental restoration and integrity, but may also galvanize agriculture into a more profitable future. Ultimately, with environmental degradation and natural resource depletion, there will be "no alternative to sustainable development."[13] While the creative destruction of the nation's economy and environment under the neoliberally controlled Farm Bill proceeds, those agribusinesses focusing on sustainable growth will have a competitive advantage if they foster sustainable innovation, thereby reducing costs and maximizing output.[14]

---

[11] Id. at 61.
[12] Id.
[13] Id.
[14] Id.

## 5.1.2   Food regulation in practice: Agent orange corn

Before embarking upon a deeper examination of comparative GMO regulation, one must understand the current food system in action. In this section, agent orange–resistant GMOs illustrate the basics of food regulation in the United States, from where much of the proliferation of GMOs springs. Specifically, agent orange corn and soybeans, Dow's 2,4-D herbicide-resistant, genetically engineered (GE) crops (hereinafter AOC) illustrate how US agencies regulate the approval process of GMOs under the Coordinated Framework.

Generally, the three agencies overseeing GMOs are the Food and Drug Administration (FDA), the Environmental Protection Agency (EPA), and the US Department of Agriculture's (USDA)[15] subdivision, Animal and Plant Health Inspection Service (APHIS).[16] First, the FDA is responsible for ensuring the safety and proper labeling of all GE plant–derived food and feed, which are not generally recognized as safe. The FDA thereby enforces the Federal Food, Drug, and Cosmetic Act (FFDCA).[17] Second, under the Federal Insecticide, Fungicide, and Rodenticide Act (FIFRA),[18] the EPA regulates the sale, distribution, and use of pesticides in order to protect health and the environment from unreasonable harm through a pesticide registration process, including plant-incorporated protectants (PIP) made by GE plants,[19] such as agent orange corn (see also Section 4.1). Third, USDA-APHIS protects plants and agriculture pursuant to the Plant Protection Act (PPA).[20] Additionally, APHIS and the EPA act under the National Environmental Policy Act (NEPA)[21] and the Administrative Procedure Act (APA).[22] In sum, the FDA asks whether GMOs are safe to eat, the USDA asks whether they are safe to plant, and the EPA asks whether they are safe for the environment.

---

[15] Emily Marden, Risk and regulation: U.S. regulatory policy on genetically modified food and agriculture, 44 *B.C. L. Rev.* 733, 767 (2003).

[16] USDA, Coordinated Framework, https://www.aphis.usda.gov/aphis/ourfocus/ biotechnology/sa_regulations/ct_agency_framework_roles.

[17] FFDCA, 21 U.S.C. §§ 301–399.

[18] FIFRA, 7 U.S.C. §§ 136–136y.

[19] Marden, *supra* note 1 at 767; USDA, *supra* note 2.; FDA, Food from Genetically Engineered Plants, https://www.fda.gov/Food/IngredientsPackagingLabeling/ GEPlants/default.htm.

[20] Marden, *supra* note 1, at 767; Draft Environmental Impact Statement, Dow AgroSciences Petitions for Determinations of Nonregulated Status for 2,4-D-Resistant Corn and Soybean Varieties, page at i (2013), https://www.aphis.usda.gov/brs/aphisdocs/24d_deis. pdf; PPA, 7 U.S.C. §§ 7701–7772 and 7 C.F.R § 340 *et seq.*

[21] NEPA, 42 U.S.C. § 4321 *et seq.*

[22] APA, 5 U.S.C. § 551 *et seq.*

These agencies work together to determine the safety of GMOs by field-testing new transgenic plants,[23] or by registering, labeling, and deregulating them, thereby monitoring the industry's marketing of GMOs. Under the Coordinated Framework, three principles guide the agencies in their GMO regulation: (1) a product-based approach focusing regulatory attention on the end product,[24] "(2) treatment of GM products as being on a continuum with other agricultural innovations, and (3) the position that regulatory action should be based on demonstrable 'scientific risk' rather than precaution."[25] The agencies and laws that comprise the Coordinated Framework are an example of the complex, legal framework of our food system in that they apply these principles through the bouquet of the above federal laws, without a streamlined consensus of the GMO risk assessment scope prior to approval.

Specifically, in the approval of Dow's AOC, this fragmented executive oversight illustrates how dangerous substances enter the US food system after passing through agency regulation (see also Chapter 2). Initially, APHIS[26] prepared an environmental assessment (EA), under NEPA, finding that Dow's products could significantly affect the environment. Therefore, APHIS followed up with an environmental impact statement (EIS).[27] Ultimately, APHIS concluded that the GE organism does not pose any plant pest risk and issued a regulatory decision of non-regulated status. Before APHIS completed the EIS in 2013, the EPA had registered and re-registered 2,4-D in 2005 because the EPA has not made a common mechanism of toxicity finding as to 2,4-D.[28] The USDA and the FDA also continue to allow the marketing of AOC.

In conclusion, the regulatory process for GMOs is intricate, complex, and convoluted. Some regulatory oversight and laws overlap or cancel each other out, while huge gaps persist within the system. For example, GE food labeling laws are underdeveloped because none of the federal agencies have decisive authority to require rigorous labeling. As Emily Marden has pointed out in her *Boston College Law Review* article, "the U.S. approach to GM foods has helped the industry grow"[29]—but industrial growth is not the point of food systems! GM foods, rather than feeding the masses, have poisoned human health by promoting pesticide-infested processed fast food, polluted the environment through pesticide-intensive

[23] Marden, *supra* note 1, at 774.

[24] Id., at 740.

[25] Id., at 784.

[26] Dow's Draft Environmental Impact Statement, at i (2013), https://www.aphis.usda.gov/brs/aphisdocs/24d_deis.pdf.

[27] Id. at iv–v, 9, 15–23.

[28] EPA, Reregistration Eligibility Decision for 2,4-D, http://archive.epa.gov/pesticides/reregistration/web/pdf/24d_red.pdf.

[29] Marden, *supra* note 1, at 735.

GE monocultures, and assaulted animal welfare through industrial meat production, where the animals are fed with GE feed. From a food ethics perspective, GMO patenting facilitates biopiracy and threatens biodiversity globally. The effects are the much-lamented losses of food safety, security, and sovereignty (see also Chapters 1 and 3).

### 5.1.3    GMOs labeling: Is there any regulation at all?

Labeling is a point of attack where the trivialization of GMOs through their proliferation may be stopped, thereby interrupting the cycle depicted in Figure 2.1. Governments are empowered to step in and shift the exercise of their influence away from serving the industry to protecting consumers and the environment. However, as the following FDA guidance illustrates, this is rarely the case.

In November 2015, the FDA issued the industry guidance on "Voluntary Labeling Indicating Whether Foods Have or Have Not Been Derived from Genetically Engineered Plants"[30] (GE foods). This guidance instructs the food industry to label GE foods if they "have characteristics that are materially different from those of comparable foods,"[31] such as "soybean oil containing higher levels of oleic acid than conventional soybean oil [which] must be labeled 'high oleic soybean oil.'"[32] GE food producers are, however, free to decide whether to label those types of foods under the condition that the labels are not misleading.

Voluntary GE food labeling is controversial because it empowers the industry and those with a stake in selling GE foods to decide when to give consumers the power to identify and, thereby, possibly avoid these products. According to a national poll, 89% of Americans support mandatory GE food labeling,[33] but the FDA's guidance makes it voluntary and difficult for consumers to know what they are purchasing and when they are eating GE foods. Instead, the guidance allows non-GE foods to be marked, favoring consumers who actively avoid GE foods. Consequently, this guidance waters down consumer rights where many are aware that

---

[30] FDA, Guidance for Industry: Voluntary Labeling Indicating Whether Foods Have or Have Not Been Derived from Genetically Engineered Plants (Nov. 2015), http://www.fda.gov/Food/GuidanceRegulation/GuidanceDocumentsRegulatoryInformation/ucm059098.htm.

[31] Id.

[32] Maggie Fox, There's No Need to Label GMO Plants, FDA Says, *NBC News* (Nov. 23, 2015), http://www.nbcnews.com/health/health-news/theres-no-need-label-gmo-plants-fda-says-n468301.

[33] Center for Food Safety, New Poll: Nearly Nine in 10 Americans Want Labels on GMO Food (Dec. 2, 2015), http://www.centerforfoodsafety.org/press-releases/4150/new-poll-nearly-nine-in-10-americans-want-labels-on-gmo-food.

GE foods may be unsafe, nutrient depleted, flavorless, substitute[34] imitations of foods, which fail the essential purpose of feeding people and cause severe environmental risks.[35] However, to others, GE foods allegedly reduce production costs and stabilize the large production of commodity crops to feed the growing population demands.[36] The latter may support GE foods, regardless of whether they are labeled as such.

On the national level, the affected parties range from consumers who buy and eat GE food, to farmers who grow GE foods, processors and producers of such foods or foods with ingredients that are GE, marketing and advertising companies promoting GE foods, including restaurants, grocery stores, and retailers offering GE foods. Moreover, the FDA, the courts, state legislatures deferring to the guidance, and policy makers are affected by the guidance in being bound by the FDA's interpretation of the FFDCA. On an international level, any party involved in the import and trade of GE foods from the United States is affected, including major trading partners in the EU and China,[37] where GE foods are opposed and embargoed.

Generally, the FFDCA empowers the FDA to oversee GE food labeling and prohibits misbranded foods from entering into interstate commerce.[38,39] According to the FDA, genetic engineering refers to the method of deriving foods, which is not material information that needs to be disclosed on labels.[40] In *Alliance for Bio-Integrity v. Shalala*, this deference to the FDA's interpretation of the FFDCA's provisions on GE food labeling was upheld.[41]

---

[34] 21 C.F.R. § 101.13(d)(1) defines "[a] 'substitute' food [as] one that may be used interchangeably with another food that it resembles, i.e., that it is organoleptically, physically, and functionally (including shelf life) similar to, and that it is not nutritionally inferior to unless it is labeled as an 'imitation.'"

[35] See generally, Andrew Kimbrell (Ed.), *The Fatal Harvest Reader*. Island Press (2002).

[36] Id.

[37] The Center for Food Safety reported that, "According to the National Grain and Feed Association, corn growers in the U.S. lost from $1 to 3 billion in revenue last year after China rejected nearly 1.5 million metric tons of U.S. corn due to contamination with a GE variety, developed by Syngenta, that China has not approved for import. China, the 3rd largest U.S. corn buyer, began importing corn from America's chief corn export competitor, Brazil, to make up the shortfall, an example of how lax U.S. policies on GE crops harms American agriculture." Center for Food Safety, U.S. Contamination Episodes Concerning Genetically Engineered Crops (Jul. 13, 2015), http://www.centerforfoodsafety.org/fact-sheets/3984/us-contamination-episodes-concerning-genetically-engineered-crops.

[38] Under Section 403(a)(1) of the FFDCA, a food is misbranded if its labeling is false or misleading in any particular. 21 U.S.C. § 343(a)(1). Section 201(n) of the FD&C Act (21 U.S.C. § 321(n)) provides that labeling is misleading if, among other things, it fails to reveal facts that are material in light of representations made or suggested in the labeling, or material with respect to consequences that may result from the use of the food to which the labeling relates under the conditions of use prescribed in the labeling, or under such conditions of use as are customary or usual. 21 U.S.C. § 321(n).

[39] 21 U.S.C. § 331(a).

[40] FDA Guidance, *supra* note 1 (internal citations omitted).

[41] Alliance for Bio-Integrity v. Shalala, 116 F. Supp. 2d 166, 178–79 (D.D.C. 2000).

As a result of this "regulatory ambiguity, consumer claims in litigation concerning GE food often focus on allegedly misleading or deceptive terms on the label when the food contains GE ingredients."[42] Thus, the resulting "inconsistency has created further ambiguity concerning the broader issue of when courts should defer to the FDA's expertise if the FDA has repeatedly declined to take action on a particular regulatory issue."[43]

Instead of monitoring GE foods on behalf of consumer protection, the FDA misses the point of why GE foods should be labeled, deferring to the industry—those whose goal is to sell and capitalize on GE foods—to "voluntarily label their foods with information … as long as such information is truthful and not misleading."[44] It is not the truthfulness of the presence or absence of GE foods in a certain product that matters, but the proliferation of GE foods as a whole and the fact that GE foods are not equivalent to non-GE varieties.[45] With this guidance, the agency blatantly ignores the public health and environmental risks of GE foods and dismisses the vast amount of scientific publications that warn against GE foods. The FDA is lulled in the pseudo-science calling GE foods safe.[46]

In sum, this guidance substantially favors the industrialization of agriculture by setting GE plant varieties on an equal footing with non-GE varieties and by stripping consumers of the right to know whether their food is GE. Therefore, the FDA's guidance misses the point of GE food labeling, and the FDA fails its mandate to act *for the people* as an executive agency. This case study shows how the industry lobby, political pressure, and economic distortions (see Figure 2.1) shape the food system into a permissive network serving GMOs proliferation by trivializing their risks.

Shifting one's perspective and considering the industry's counterargument, however, reveals the fallacies of the position. The industry argues that the FDA's guidance conforms to its mission to act for the benefit of the people in making GE food labeling voluntary for the industry. Notably, the FDA's agricultural policy most effectively defends its alleged mandate "when a food product is mislabeled, presents a health risk because of contamination, or has caused an outbreak of illness."[47] The FDA has a fast

---

[42] Emily M. Lanza, Legal Issues with Federal Labeling of Genetically Engineered Food: In Brief, Congressional Research Service (Sept. 22, 2015), https://www.fas.org/sgp/crs/misc/R43705.pdf.

[43] Id.

[44] Fox, *supra* note 3.

[45] Hollingworth RM, Bjeldanes LF, Bolger M, Kimber I, Meade BJ, Taylor SL, and Wallace KB, Society of Toxicology ad hoc Working Group. The Safety of genetically modified foods produced through biotechnology, *Toxicol. Sci.* 71 (1): 2–8 (2003).

[46] See generally, Sheldon Krimsky, Jeremy Gruber, and Ralph Nader (Eds.), *The GMO Deception*, (2014).

[47] FDA, Recalls, Outbreaks & Emergencies, http://www.fda.gov/Food/RecallsOutbreaksEmergencies/default.htm.

response system that quickly processes investigations into food safety, if in doubt, and updates food recalls regularly on its website and via an app,[48] thereby effectively alerting consumers to food outbreaks—or so the FDA claims. Especially in times of need, people can turn to the FDA, who, "along with the U.S. Centers for Disease Control and Prevention (CDC), provides information on food safety and disposal during public health emergencies, such as floods, hurricanes, earthquakes, and other natural disasters."[49] Warning consumers of GE foods, which have been marketed and consumed without additional labeling for nearly two decades, is outside of the FDA's purview because those are not misleading. In short, the FDA and the CDC do more to protect producers than consumers.

As with the guidance on GE food labeling, the FDA clearly outlines which foods are considered misleading under the FFDCA: if the label "fails to reveal facts that are material in light of representations made or suggested in the labeling, or material with respect to consequences that may result from the use of the food to which the labeling relates...."[50] Thus, if any foods are mislabeled, the FDA prohibits their introduction into interstate commerce. For those trying to apply the FDA's guidance, however, it is everything but clear.

In the guidance in question here, the FDA clearly explains when a food would be mislabeled as being non-GE, where "[m]anufacturers often voluntarily provide information on their labels beyond the information required by the [FFDCA] or FDA regulations. Their reasons for doing so may have to do with marketing or providing information of specific interest to their customers."[51] Consequently, the FDA monitors food labels and ensures that mislabeled products are taken off the market. Removing such products ultimately protects consumers from being exposed to foods that the FDA deems unsafe—except for those numerous consumers who have already suffered slowly progressing illnesses as a result of the FDA's flawed system of safeguarding consumer interests.

The industry may, of course, label foods as non-GE or as containing GE ingredients, pursuant to the FFDCA. As long as the industry, however, complies with the FFDCA and other rules and regulations within the FDA's jurisdiction, such as the Food Safety Modernization Act (FSMA), the FDA has no reason to further restrict the market freedom of food producers and the industry. For instance, the FDA enforces the FSMA to

---

[48] FDA, Recalls, Market Withdrawals, & Safety Alerts, http://www.fda.gov/Safety/Recalls/default.htm.

[49] Id.

[50] Guidance (citing 21 U.S.C. § 321(n)).

[51] FDA, Guidance for Industry: Voluntary Labeling Indicating Whether Foods Have or Have Not Been Derived from Genetically Engineered Plants (Nov. 2015), http://www.fda.gov/Food/GuidanceRegulation/GuidanceDocumentsRegulatoryInformation/ucm059098.htm.

"ensure the U.S. food supply is safe by shifting the focus from responding to contamination to preventing it."[52] In other words, the guidance to the industry in question merely clarifies what the industry may label as GE or non-GE within the FFDCA's framework. The FDA is, according to its own whitewashed lies, acting in the public good while simultaneously clarifying how the industry may comply with the FFDCA if producers choose to label GE foods at all. Finally, the FDA's guidance, here, allegedly helps both consumers and the industry to understand the meaning of "misleading" food labels, preventing confusion. It is now known that none of this is true.

Thus, after playing devil's advocate and examining the alternative viewpoint, one might ask whether there is a middle ground in this discussion? Unfortunately, there is none, because foods are either GE or they are not. It would be logical to label them as such instead of hiding this information. Failing to require GE food labeling, therefore, is an abuse of discretion by relying on an outdated definition of "material,"[53] allowing producers to merge their likely harmful products with non-GE varieties. Additionally, calling GE crops, for instance, the euphemism "conventional" instead of the truth, "hazardous," is another way that the industry deceives consumers. Consequently, the FDA's failure to update its regulations and definitions evidences the pro-industry attitude, which strips consumers of their right to choose what to eat and even confuses scientists.[54]

Genetically engineering plants for food, for example, means that those plants are altered genetically to either resist high levels of herbicides, to produce pesticides, or to thrive with a specific chemical fertilizer use—or all three.[55] The goal is generally not to feed the world, even though it has been propagated as a falsely altruistic goal of the industry.[56] The FDA's explanation in the guidance, for instance, furthers the industry's explanations, rather than the scientifically correct one by stating that,

[52] FDA, Food Safety Modernization Act (FSMA), http://www.fda.gov/Food/GuidanceRegulation/FSMA/default.htm.

[53] "Material" is limited to alterations that can be sensed, that is, tasted, seen, or smelled, which is an unfair limitation on the importance of other food traits, such as nutritious value. See also the Guidance (considering a food attribute material "in cases where a consumer may assume that a food, because of its similarity to another food, has nutritional, organoleptic (e.g., taste, smell, or texture), or functional characteristics of the food it resembles when in fact it does not (e.g., a statement that reduced fat margarine is not suitable for frying (21 CFR 101.13(d)(1))).")

[54] Sheldon Krimsky, Jeremy Gruber, and Ralph Nader, *The GMO Deception* (2014).

[55] See generally, World Health Organization, Food Safety, http://www.who.int/foodsafety/areas_work/food-technology/faq-genetically-modified-food/en/.

[56] Brooke Borel, Core Truths: 10 Common GMO Claims Debunked, Popular Science, http://www.popsci.com/article/science/core-truths-10-common-gmo-claims-debunked.

> the term "genetically modified" can encompass
> any alteration to the genetic composition of a plant,
> including alterations achieved through traditional
> hybridization or breeding techniques, that term
> could apply to most cultivated food crops since
> most food crops are the product of selective breed-
> ing. An example of a food that is derived from a
> plant that has not been subject to any form of selec-
> tive breeding might be berries collected from wild
> plant varieties.[57]

By analogy, the coexistence of conventional agriculture based on GE plants, on the one hand, and organic farming, largely excluding GE plants, on the other hand, is an industry-created myth. GE plants do not merely coexist with wild-type plants or selectively bred counterparts. GE varieties either extinguish and outcompete them or infiltrate the wild types with their engineered DNA through cross-pollination and genetic drift.[58] Even the guidance acknowledges that "[f]oods that comply with ... USDA organic regulations ... would meet criteria to be labeled as not produced or handled using bioengineering."[59] The *Seattle Times* published a nice infographic that illustrates the differences between genetic engineering and plant breeding, a distinction that the FDA got wrong in its guidance.

In conclusion, there is no middle ground because the GE varieties exist at the expense of the wild types. As the Center for Food Safety's Colin O'Neil explains in *The GMO Deception*, "unlabeled GE foods are misleading not only because they contain unperceivable genetic and molecular changes to food, but also because they have unknown and *undisclosed* risks"[60] (emphasis added). In fact, the FDA conducted no independent food safety tests on GE foods and does not affirm their safety.[61] Thus, the FDA leaves unsafe foods, as various scientific publications have proven, unlabeled and untested, and confound consumers into thinking they are alright to eat.[62] This is not what an agency does on behalf of the public. It is an endorsement of the industry.

The standardization of GMOs in our food system reduces market barriers (see Figure 2.1, top left quadrant) and promotes the proliferation of

---

[57] FDA, Food Safety Modernization Act (FSMA), http://www.fda.gov/Food/GuidanceRegulation/FSMA/default.htm.

[58] Textbox: Cross-contamination, genetic drift, and the question of GMO co-existence with non-GM crops, in *International Food Law and Policy* (Gabriela Steier and Kiran Patel, Eds.). Springer International (2017).

[59] Id. (citing 7 CFR § 205).

[60] Krimsky, *supra* note XX.

[61] Id.

[62] Id.

GMOs. The FDA would be in a position to exercise its mandate and protect the public from GMO foods, thereby stopping or slowing the industry's trivialization of the risks that GMOs pose to consumers and the environment. Labeling GMOs would be an excellent method to halt the trivialization, thereby interrupting the cycle that feeds into the proliferation of GMOs, which takes over the food systems, as explained in Chapter 1. As such, labeling is another point of attack where the trivialization of GMOs through their proliferation may be stopped.

## 5.1.4    EFSA versus FDA: Better safe than sorry!

Taking a perspective-shifting approach, as explained in Chapters 1 and 2, helps to compare and find solutions for shared or analogous problems. As such, contrasting two vastly different frameworks for the regulation of GMOs is fertile ground to learn from the intricacies of both systems and to extract valuable insights for the future. Specifically, the EU and US approaches to regulating GMOs differ greatly and the EU better protects its citizens from unsafe GE foods. According to the precautionary principle,[63] the EU Food Safety Authority (EFSA) prevents GE foods unless proven safe. In contrast, the US policy favors biotechnology and allows GMOs unless proven unsafe, thereby giving the industry the benefit of the doubt. Thus, the US approach includes a cost-benefit analysis "and avoids hindering safe innovations."[64] However, the safety of GMOs is doubtful according to over 300 renowned scientists reporting that "the scarcity and contradictory nature of the scientific evidence published to date [which] prevents conclusive claims of safety, or of lack of safety, of GMOs."[65] Nonetheless, the United States continues to approve GMOs for the probable reason of facilitating economic growth and innovation.[66]

US deregulation of GMOs extends beyond national borders, where biotechnology companies aggressively promote[67] and patent GMOs abroad, thereby jeopardizing food security, resilience to climate change, and sustainability in trading partners importing American GMOs.[68] In fact, the biotech companies capitalizing on GMOs create many of the

---

[63] Eur-Lex, Precautionary Principle, http://eur-lex.europa.eu/legal-content/EN/TXT/?uri=URISERV:l32042.

[64] Marden, *supra* note 1, at 741–742.

[65] Angelika Hilbeck et al., No scientific consensus on GMO safety, 27 *Environmental Sciences Europe* 4, 1–6, 1 (Springer 2015), http://www.enveurope.com/content/pdf/s12302-014-0034-1.pdf.

[66] See generally, Marden, *supra* note 1.

[67] Food and Water Watch (FWW), Greenwashing GE Crops, http://www.foodandwaterwatch.org/insight/greenwashing-ge-crops.

[68] Food and Water Watch, GMOs: Promoting GMOs Abroad, http://www.foodandwaterwatch.org/problems/gmos.

problems that they claim to solve.[69] Thus, as Professor Krimsky from Tufts University points out, "'scientific consensus on the safety and agricultural value of GM crops'—is laying the basis for major, uncontrollable blowbacks."[70] Moreover, GMOs need not be labeled in the United States, where the so-called Deny Americans the Right to Know (DARK) Act, that is, the Safe and Accurate Food Labeling Act of 2015, keeps consumers in the dark about whether their foods are GE.[71] In contrast, the EU mandates GMO labeling, thereby increasing the transparency and traceability of GMOs.[72] Thus, the EU empowers consumers to avoid GMOs for food safety reasons.

The resulting sliding scale across the Atlantic has given rise to the hotly debated Transatlantic Trade and Investment Partnership (TTIP) discussed in Chapter 2. One of the main points of discontent are genetically altered animals, such as the recently approved GE salmon from the United States, which would be forced upon the EU if the TTIP came into being despite consumer resistance. Section 5.2 explains the controversy based on the frankenfish case study.

## 5.2   *Transgenic animals*

Until recently, only patented plants existed on the market for food and feed. Despite all of the concerns over their safety for consumers, animals, and the environment, GE plants remained fairly silent and faceless problems. On the imaginary and hypothetical chessboard introduced in Chapter 1, GE plants were the runners or pawns, which could somehow be pushed around and used interchangeably. They were numerous front-runners of the GMO battle. Transgenic animals, however, are rooks and knights of the chess game: they make or break the match and give a face to the GMOs debate.

### 5.2.1   *Special case of GE salmon: An analysis*
### *of the frankenfish debate*

On November 19, 2015, the FDA announced its long-awaited and often feared decision regarding AquAdvantage salmon (AAS), the first GE

[69] FWW, *supra* note 11.

[70] Sheldon Krimsky and Jeremy Gruber (Eds.), *The GMO Deception* (2014). Kindle locations 331–332.

[71] Anna Roth, 5 Things to know about the 'DARK Act,' *CivilEats* (Jul. 20, 2015), http://civile-ats.com/2015/07/20/5-things-to-know-about-the-dark-act/.

[72] European Commission, Plants: Traceability and Labeling, http://ec.europa.eu/food/plant/gmo/traceability_labelling/index_en.htm.

animal intended and approved as food.[73] The FDA approved AquaBounty Technologies' new animal drug application (NADA) for the salmon, FDA, *Draft Guidance for Industry: Voluntary Labeling Indicating Whether Food Has or Has Not Been Derived from Genetically Engineered Atlantic Salmon* (hereinafter frankenfish guidance), concluding that it complies with the statutory requirements under the FFDCA. In particular, the FDA determined that the salmon's rDNA construct is safe for the fish itself, that the fish reaches market size more quickly than non-GE, farm-raised Atlantic salmon (as AquaBounty claimed), and that food from the fish is safe to eat and as nutritious as its non-GE counterpart. The FDA also concluded that labeling of the GE salmon is not required. How dangerous!

The frankenfish guidance is a textbook example of how the trivialization of the risks of GMOs leads to their proliferation. Here, the industry has managed to trivialize and water down the risks to the point of FDA approval, notwithstanding the low thresholds of statutory protection under the FFDCA—a result that would likely have been the opposite in the EU under the EFSA's purview. With FDA approval, there is little to stand in the way of vast proliferation by marketing frankenfish in the United States and introducing it into international trade, especially if the TTIP comes to pass. Once again, Figure 2.1 illustrates in its top half how the cycle of GMOs in the food system continues. Here, the point to disrupt the cycle is through consumer actions. Consumers should request that the FDA commissioner revokes the recent approval of AAS under Section 512(c)(1) of the FFDCA[74] to protect consumers and that the FDA prepares an EIS, as required by the NEPA[75] and the Endangered Species Act (ESA)[76] and in compliance with the FSMA to protect the environment and the future of our food system.

As drafted, the FDA's guidance fails the agency's mandate to act in the public interest and overlooks the underlying public health, food safety, and environmental threats from GE salmon. The FDA's failure to require GE food labeling risks that GE foods "will induce subtle changes in long-term health and nutritional quality, increase food allergies, incentivize non-sustainable farming practices, create dependency on chemical inputs, justify a lack of transparency in evaluating food quality and safety, or

---

[73] Sections of this chapter were drafted simultaneously and may overlap with Nicole E. Negowetti and Gabriela Steier, The "Frankenfish" Debate: Approval and Labeling of Genetically Engineered Salmon, *American Bar Association Environmental Disclosure Committee Newsletter* 12–18 (Nov. 2016), http://www.americanbar.org/content/dam/aba/publications/nr_newsletters/ed/201611-ed-gmo-joint.authcheckdam.pdf.

[74] FDA, AquAdvantage Salmon Approval Letter and Appendix (Nov. 19, 2015), http://www.fda.gov/AnimalVeterinary/DevelopmentApprovalProcess/GeneticEngineering/GeneticallyEngineeredAnimals/ucm466214.htm (*citing* 21 U.S.C. 360b(a)(1))).

[75] 42 U.S.C. § 4321 *et seq.*

[76] 16 U.S.C.A. § 1531 *et seq.*

transform farming practices into a political economy resembling serf-dom…."[77] Thus, the FDA's guidance is a race to the bottom for American GE food regulation and lacks proper risk assessment. Voluntary labeling of GE salmon deprives consumers of their right to know which foods are GE so they can avoid them. Finally, the FDA's finding that GE salmon is not materially different from non-GE varieties lacks proper risk assessment.

It is imperative that the FDA mandate labeling of GE salmon to protect consumers and the environment from the detrimental effects of GMOs. The FDA's draft guidance in 80 FR 73193 (November 24, 2015) relates to the FDA's November 19, 2015, approval of a NADA for an rDNA construct in a line of farm-raised GE salmon.[78] As stated, the FDA intends the guid-ance "to assist food manufacturers that wish to voluntarily label their food products or ingredients … derived from Atlantic salmon as either containing or not containing products from genetically engineered (GE) Atlantic salmon."[79] The FDA's "concern … is that such voluntary label-ing be truthful and not misleading,"[80] and the FDA has the discretion to regulate AquAdvantage salmon accordingly and ensure that it complies with the FFDCA.[81] Thus, in a 1992 "Statement of Policy: Foods Derived from New Plant Varieties," the FDA erroneously interpreted the FFDCA, concluding that GE foods, as compared to non-GE varieties, do not differ in "any meaningful or uniform way, or that, as a class, foods developed by the new techniques present any different or greater safety concern than foods developed by non-GE plant breeding."[82] In applying these princi-ples, the FDA determined that AAS is substantially equivalent to Atlantic salmon.[83] This chapter continues to explain how the trivialization of the risks associated with the approval of AAS proliferates GMOs and jeopar-dizes the food systems along with the pillars it relies on.

---

[77] Colin O'Neil, Consumers Call on FDA to LABEL GMO Foods, in *The GMO Deception* (Sheldon Krimsky, Jeremy Gruber, and Ralph Nader, Eds.) (2014). at Kindle location 1472. *The GMO Deception* is a collection of essays by various experts and academics, many of whom are affiliated to advocacy groups opposing the genetic engineering of plants and animals for food. The arguments outlined in this memorandum are based on various issues raised in this collection of essays to provide a comprehensive analysis of the spec-trum of reasons opposing GE salmon.

[78] FDA, Draft Guidance for Industry: Voluntary Labeling Indicating Whether Food Has or Has Not Been Derived From Genetically Engineered Atlantic Salmon (Nov. 2015), http://www.fda.gov/Food/GuidanceRegulation/GuidanceDocumentsRegulatoryInformation/ucm469802.htm.

[79] Id.

[80] Id.

[81] FDA, AquAdvantage Salmon Approval Letter and Appendix (Nov. 19, 2015), http://www.fda.gov/AnimalVeterinary/DevelopmentApprovalProcess/GeneticEngineering/GeneticallyEngineeredAnimals/ucm466214.htm (*citing* 21 U.S.C. 360b(a)(1)).

[82] Id. (internal citations omitted).

[83] FDA, *supra* note 3.

## 5.2.2   Unlabeled GE salmon poses risks to consumers: Nutritious inferiority, allergenicity, and environmental harm

Food labels work like a passport for foods, a common form of identification that allows categorization. Moreover, labels give consumers power to select foods they wish to avoid, as explained in Section 5.1.3. Labeling GE salmon speaks to the standard of the fish's identity pursuant to the FFDCA's labeling requirements. Thus, failure to identify substances added to foods, such as colorings,[84] already constitutes misbranding and results in deception or unfair competition.[85] The addition of the rDNA construct in AAS can be construed as an additive, thereby triggering the FFDCA's labeling requirements.[86] Moreover, given the scientific uncertainty regarding GE salmon food safety, the FFDCA provisions regarding unsafe food additives likely apply and mandate labeling because there are no current "conditions under which such additive may be safely used."[87] Unless the FDA exempts GE salmon under 21 U.S.C.A. § 379e(f) consistent with public health, for which the FDA lacks the scientific support, the guidance in question contravenes the FFDCA's labeling requirements.

GE salmon's nutritional profile differs from that of non-GE varieties and is, therefore, materially different. AquaBounty's data submitted to the FDA revealed that AAS "contains less healthy fatty acids than other farmed salmon and far less healthy fatty acids than wild salmon." Therefore, the FDA should re-examine its "claims that the omega-3 to omega-6 fatty acid ratio in [ASS] ... is similar to the ratios found in scientific literature for farmed Atlantic salmon."[88] Additionally, the mineral and vitamin levels in GE salmon are also lower, "by more than 10%"[89] and AAS "has lower levels of every essential amino acid tested and nearly 25% less folic acid and vitamin C."[90] In fact, AAS is fattier and less nutritious as a result of its genetic modification.[91] These material differences between GE salmon and its non-GE counterpart should be labeled to help consumers decipher when they are supporting the proliferation of this frankenfish with their grocery purchases.

---

[84] 21 U.S.C.A. § 343(g), (i) (West).
[85] 21 U.S.C.A. § 343 (West).
[86] 21 U.S.C.A. §§ 379e, 343(m).
[87] 21 U.S.C.A. § 379e(a)(1)(A).
[88] Colin O'Neil, Consumers Call on FDA to LABEL GMO Foods, in *The GMO Deception* (Sheldon Krimsky, Jeremy Gruber, and Ralph Nader, Eds.) (2014) (at Kindle location 1487) ("In fact, the ratio for [AAS] ... is nearly 15 percent less than the recorded ratio for conventionally farmed Atlantic salmon and 65 percent less than wild salmon.")
[89] Id. at Kindle location 1487.
[90] Id.
[91] Id.

GE foods should also be labeled to warn allergy sufferers and diabetics.[92] According to a recent study, "Atlantic salmon in itself is a known allergenic food, thus it is probable that" AAS is, too.[93] It should be labeled to warn allergy sufferers because "consumers actually nee[d] information about genetic modifications ... as the FDA itself has recognized in the Federal Register...."[94] Other food safety concerns with AAS involve the increased hormone content that might affect diabetics. Scientists warn that, in AAS, "the higher concentration of insulin-like growth factor-1 ... represents a critical issue."[95] They emphasize that "it is unsettling that the FDA hasn't conducted a complete analysis..."[96] prior to approving AAS and that it will be marketed unlabeled and without further independent studies to reveal other AAS food safety concerns.

Transgenic animals are not regulated as food, despite their use as such. Alarmingly,

> the transgenic animals are regulated as veterinary drugs and not as food, so that AquaBounty can effectively market the GE salmon animal drug as food but evade the appropriate regulation. Indeed, according to US regulatory processes, the transfer of genetic information can be viewed as a way to deliver a drug ... to the tissue of the animal[97]

thereby potentially mass-medicating consumers against their consent because, without a label, they do not know when they are buying AAS. Consequently, in addition to nutrient deficiency, allergies, and diabetes, AAS could trigger untraceable drug interactions in some individuals, especially if it remains unlabeled and untraceable for people with sensitivities—issues ripe for FDA regulation through mandatory GE salmon labeling.

GE salmon also poses uncalculated risks for environmental harm and could endanger ecosystems on a large scale. AAS production is energy and resource intensive, unsustainable, and potentially dangerous. Through advertisements and selective data submissions to the FDA, AquaBounty has green-washed AAS production, but the massive land-based fish factories have a substantially larger environmental

---

[92] Id. at Kindle location 1513.
[93] Alice Benessia and Giuseppe Barbiero, The impact of genetically modified salmon: From risk assessment to quality evaluation, *Visions for Sustainability* 3, 35–61 (2015), http://www.iris-sostenibilita.net/public/vfs/pdf/VFS-20150001095.pdf.
[94] *The GMO Deception, supra* note 13, at Kindle location 1602.
[95] Benessia and Barbiero, *supra* note 18.
[96] Id.
[97] Id. at 48.

footprint than naturally occurring salmon or even salmon farms. The land-based tanks, for instance, require oxygenation, sludge and waste removal, mechanical filters to clean the fish tanks, electric water pumps,[98] and other processes that, in a natural marine environment, are part of a functioning, self-replenishing, and balanced oxygen cycle, whereby water plants and algae produce oxygen. The land-based, artificially oxygenated and cleaned salmon factories endanger the natural habitats of other species, alter land use, and promote unsustainable aquaculture. Unless given a choice through GE salmon labels, consumers will inadvertently be forced to finance threats to their environment through AAS purchases.

## 5.2.3   *Voluntary labeling of GE salmon and consumers' right to know*

Seemingly prescriptive in nature, this subsection provides a framework for how the FDA and consumer could fight the proliferation of transgenic animals, spearheaded by AAS, to protect the food system from industry takeover. Specifically, consumers are interested in how their food is made and its material label information. The guidance describes technically truthful yet incomplete voluntary labels in contravention of public interest and should, therefore, be revised to include all the information that consumers need to make educated choices.[99] According to a nationwide poll, 95% of respondents want to know whether their food has been genetically modified.[100] At its core, this is a right to make autonomous and informed decisions, rather than being subjected to broad-sweeping regulatory inaction at the expense of consumer interests. Therefore, the FDA should inform consumers—through complete labels—and revise the guidance with mandatory identifications of AAS.

The FDA should correct its current interpretation of materiality in food labels,[101] because labels should facilitate communication between seller and buyer. In the guidance, the FDA erroneously concludes that genetic engineering of salmon is not "material," although it is, in fact, synonymous with "important" or "significant," dependent on personal judgment that rests in the eye of the beholder.[102] In other words, consumers must not be deprived of the autonomy to decide for themselves whether genetic manipulation of their salmon factors in their buying and consumption decisions. Mandatory GE labeling would merely mirror other existing process labels, such as "kosher, dolphin-free, Made in America,

---

[98] AquaBounty, Sustainable, https://aquabounty.com/sustainable/.
[99] *The GMO Deception*, *supra* note 13 at 1513.
[100] Id. at Kindle location 1493.
[101] Id.
[102] Id. at 1527–1528; *Oxford English Dictionary*, "Material," http://www.oxforddictionaries.com/definition/english/material.

union-made, free-range, … irradiated, and … 'organic.'"[103] All of these labels indicate how the food product was derived, not what it contains.[104] Logically, any argument of material differences in products bearing these labels is void because kosher meat, for instance, is chemically identical to non-kosher meat.

Moreover, the current involuntary labeling weakens consumers because the FDA's guidance treats the AAS labeling akin to eco-labels, without regulatory support. Eco-labels are increasingly popular labels bearing information about the environmental friendliness of products with the objective "to harness market forces and channel them towards promoting more environmentally friendly patterns of production."[105] However, the FDA's proposed characterization of voluntary GE salmon labeling deprives consumers of the pertinent information whether the salmon they are purchasing is GE and grown in land-based tanks— important environmental considerations.

Consumers want to know what they are buying, and they have a right to know. The introduction of a bill to amend the FFDCA to require labeling of GE fish[106] further supports the observation that there is substantial demand to identify these products. According to this proposed amendment, a food is misbranded pursuant to 21 U.S.C. § 343 "[i]f it contains genetically engineered fish unless the food bears a label stating that the food contains genetically engineered fish."[107] In light of the growing demand for eco-labeling and consumer information, the FDA's foresight in mandating GE salmon labeling now would evidence the popular demand that will likely grow in the near future.

Additionally, consumers lack redressability under the FFDCA, and the guidance will force GE salmon on consumers without giving them a choice to avoid it nor to claim their rights. The consolidated decision, *In re Farm Raised Salmon Cases*, before a California Court of Appeals, explains that the FFDCA "preempts individuals' state-law claims against several grocery stores for selling artificially colored farmed salmon without disclosing to consumers the artificial coloring." In enacting FFDCA §337(a), Congress made clear its intention to preclude private enforcement of the FFDCA and that a state-law private right-of-action based on an FFDCA violation would frustrate the purposes of exclusive federal

---

[103] Id. at 1545.
[104] Id.
[105] Matthias Vogt, Environmental labeling and certification schemes: A modern way to green the world or GATT/WTO-illegal trade barrier? *Environmental Law Reporter* 33, 10522 (2003).
[106] H.R.584.IH (Feb. 6, 2013), http://thomas.loc.gov/cgi-bin/query/z?c113:H.R.+584.
[107] Id.

and state governmental prosecution of the Act."[108] Notably, prior to FDA's approval of AAS, "12 Senators led by Senator Mark Begich (D-Alaska) and 21 Representatives led by Congressmen Don Young (R-AK-01), Mike Thompson (D-CA-05) and Jared Huffman (D-CA-02) sent letters to the FDA urging it to halt its approval until their economic, regulatory and environmental concerns are addressed."[109]

However, until the FDA requires labeling of AAS, consumers will remain defenseless against the mass marketing and proliferation of GE salmon that many oppose and will be forced into all-or-nothing decisions regarding their purchase of salmon if they wish to avoid unlabeled AAS.

## 5.2.4    Opportunities for improvement from around the world: Requiring independent scientific testing to ensure food safety

The "guidanc[e] describe[s] FDA's current thinking on"[110] the topic of GE salmon and shows that the FDA's concern is on the truthful and not misleading labeling of hazardous GE salmon products. This emphasis on technicalities over broader policy goals embodies an outdated minority approach compared to other countries. Nonetheless, the current guidance empowers the FDA to revise and update its position. International treaties as potential examples for the FDA's reconsideration of its GE labeling policy follow.

First, the *Cartagena Protocol on Biosafety to the Convention on Biological Diversity* is an international agreement with 170 signatories, that "aims to ensure the safe handling, transport and use of living modified organisms … resulting from modern biotechnology that may have adverse effects on biological diversity, taking also into account risks to human health."[111] Additionally, the protocol posits that economic and trade interests should be balanced against public health and environmental considerations. Second, the *Codex Alimentarius* under the World Health Organization and the Food and Agriculture Organization of the United Nations, sets "international food standards, guidelines and codes of practice contribut[ing] to the safety, quality and fairness of this international food trade. Consumers can trust the safety and quality of the food products they buy and importers can trust that the food they ordered will be in accordance

---

[108] In *re Farm Raised Salmon Cases*, 36 ELR 20178 (Cal. App. 2d Dist., 08/31/2006), http://stevens.vermontlaw.edu:2069/litigation/36/20178/farm-raised-salmon-cases-re.

[109] Center For Food Safety, Press Release: Nearly 2 Million People Tell FDA Not To Approve GE Salmon, http://www.centerforfoodsafety.org/files/senate-to-fda-ge-salmon-42413_28714.pdf.

[110] Id.

[111] Convention on Biological Diversity, About the Protocol, https://bch.cbd.int/protocol/background/.

with their specifications."[112] Finally, the EFSA, an agency with equivalent regulatory oversight as the FDA, established a panel of experts to provide "independent scientific advice on food and feed safety, environmental risk assessment and molecular characterization/plant science" including GE animals.[113] All of these international examples have paved the way for the FDA to take a more proactive role in regulating GE salmon by balancing the public health against the industry's interests, encouraging consumer trust in FDA-regulated foods, and by requiring independent safety tests of GE salmon. Overall, the Cartagena Protocol, Codex, and the EFSA illustrate how and justify why the FDA should require mandatory labeling of AAS.

In essence, the federal US government has hesitantly approved GE salmon and concedes through recent actions that AAS should be properly labeled before being marketed to the public. For instance, Congress' December 2015 spending bill[114] "blocks any commercial sale of those salmon until the FDA finalizes its guidelines for labeling GMOs."[115] Sen. Lisa Murkowski (R-Alaska) pushed for this bill to protect the public from AAS.[116] Moreover, the FDA's recent Import Alert #99-40 "directs that during [fiscal year 2016] the FDA shall not allow the introduction or delivery for introduction into interstate commerce of any food that contains genetically engineered salmon, until FDA publishes final labeling guidelines for informing consumers of such content."[117] This lack of finality makes it possible to promote incremental or experimental shifts in policy[118] and creates a window for the FDA to introduce stricter GE labeling requirements that consider the aforementioned concerns over AAS. Thus, the FDA is empowered to mandate the labeling of GE salmon to inform consumers and to ensure traceability in interstate commerce.

[112] WHO and FAO, Codex Alimentarius, http://www.fao.org/fao-who-codexalimentarius/about-codex/en/.

[113] EFSA, Panel on Genetically Modified Organisms, http://www.efsa.europa.eu/en/panels/gmo.

[114] Fiscal Year (FY) 2016 Omnibus Appropriations Act covering the funding of the federal government during fiscal year 2016 (FY16) was signed into law by the president on December 18, 2015, becoming Public Law No: 114–113. FDA, Import Alert # 99-40 (Jan. 29, 2016), http://www.accessdata.fda.gov/cms_ia/importalert_1152.html.

[115] Allison Aubrey and Dan Charles, Follow The Money: Congress Uses Budget Bill To Rewrite Food Policies, NPR (Dec. 16, 2015), http://www.npr.org/sections/the-salt/2015/12/16/459986395/follow-the-money-congress-uses-budget-bill-to-rewrite-food-policies.

[116] Brady Dennis, FDA Bans Imports of Genetically Engineered Salmon—For Now, *Washington Post* (Jan. 29, 2016), https://www.washingtonpost.com/news/to-your-health/wp/2016/01/29/fda-bans-imports-of-genetically-engineered-salmon-for-now/.

[117] FDA, Import Alert # 99-40 (Jan. 29, 2016), http://www.accessdata.fda.gov/cms_ia/importalert_1152.html.

[118] Philip J. Harter, The Policy Process at Government Agencies 277 (undated).

Such small tweaks could, for instance, close the gaps that the industry abuses in the cycle illustrated in Figure 2.1 to seize control over executive agencies, such as the FDA, with lobbying pressure and factual distortions. Thus, the guidance is an opportunity to update the FDA's policy on GE food labeling and to reclaim regulatory control over consumer information on food labels to protect public health, food safety, and environmental integrity.

## 5.3   Summary

Using the case study of AOC, this chapter shows how GMOs are integrated into the food regulatory network in the United States. In response to all those who wonder why nothing is being done to stop the proliferation of GMOs, a brief answer would be that one needs information upon which to act, and the industry does not easily relinquish control of this information. Shifting one's perspective from the industry's release of information, one may reach the point of the public's demand for information—the right to know. This right to know embodies a forced release of information, which in food law is simply called labeling. The second part of this chapter zooms in on transgenic animals, the first GMOs that are approved in the animal kingdom: patented fish. Developing the power of the right to know and product labeling further provides information about the consumer-oriented controls to create points of attack on the proliferation of GMOs and the corresponding trivialization of the risks of a GMO-dominated food system.

# chapter six

# EU-US disputes over GMOs and the WTO biotech cases

Turning the anecdotal chessboard for the last time in this book, this chapter zooms out to the macroscopic impact that the trivialization through proliferation of genetically modified organisms (GMOs) has globally. In other words, what happens when countries disagree over the risks of GMOs? As Chapter 5 showed, there are irreconcilable differences in the regulation of GMOs between the European Union and the United States, and yet, the trade in food across the Atlantic continues. Returning to the lower right quadrant of Figure 2.1, there are market shifts resulting from GMO trade. Disputes over such trade usually go before the World Trade Organization (WTO).

Imagine the WTO as a global body of oversight dealing with most of the countries in the world to ensure that international trade runs as smoothly as possible. According to its own description, the WTO

> is the only global international organization dealing with the rules of trade between nations. At its heart are the WTO agreements, negotiated and signed by the bulk of the world's trading nations and ratified in their parliaments. The goal is to help producers of goods and services, exporters, and importers conduct their business.

The rules by which the WTO settles disputes are binding for all its 190 signatories, thereby creating a sort of international precedent[1] that does not create common law, but rather establishes how similar trade disputes may be resolved in the future. Consequently, in terms of international food trade, the WTO has the immense power to decide one case and influence all of its signatories at the same time. At the same time, this author has previously noted that,

---

[1] Gabriela Steier, The WTO's blind spot: dispute resolution in the international food industry, 11 Mayhew Hite Report 4 (Apr. 2013), http://moritzlaw.osu.edu/epub/mayhew-hite/2013/04/the-wtos-blind-spot-dispute-resolution-in-the-international-food-industry/.

> the WTO makes it clear that it does not pass judgment.
> This may be one of the problems associated with the
> WTO dispute settlement process because it does not
> help the nations in conflict find the "right" way, but
> only a way without conflict that facilitates future
> trade. Thus, the WTO does not currently set precedent
> with public policy goals – but maybe it should.

Although the details of the WTO dispute resolution process are beyond the scope of this book, what the WTO can do[2] is critical to understanding its role in enabling the proliferation of GMOs in the international market. In brief, when there is a trade problem between WTO signatories, talks and negotiations lead to an agreement and rules that are binding.[3] The WTO continues to monitor the implementation of those rules and collects and shares information for discussion in its committees that lead to a dispute settlement,[4] which resembles but is not quite like a court's decision.

Practically, after going through the WTO's dispute resolution process, governments promise to lower trade barriers; trim red tape in customs; ensure justifiable—that is, not arbitrary—restrictions on imports for health, safety, and environmental grounds; limit agricultural subsidies; and protect intellectual property.[5] Moreover, much of the WTO's work takes place "out of the limelight."[6] All of this means that the WTO has the power to promote the proliferation of GMOs or to stop it, putting the entire cycle depicted in Figure 2.1 under its jurisdiction. If the WTO were to consider the EU's decision that GMOs are unsafe until proven safe, according to the precautionary principle outlined in Chapter 4, to be arbitrary and require justification for the European embargo on GMOs from the United States, the WTO would have the power to bulldoze Europe's objections to GMOs, effectively overruling decisions issued by the EU Food Safety Authority (EFSA). Keeping this dispute resolution out of the limelight could, furthermore, keep the media in the dark and, thereby, strip consumers of a chance to appeal to the WTO to review its findings. How undemocratic! This, in fact, happened with a series of cases outlined in the following sections of this chapter.

---

[2] WTO, What the WTO can do (undated), https://www.wto.org/english/res_e/publications_e/wtocan_e.pdf.

[3] Id. at 53.

[4] Id.

[5] Id. at 52.

[6] Id. at 51.

## 6.1    Mapping the disputes: Where are the bullies?

Since the WTO dispute settlement process was instituted in 1994, a total of 623 cases have been brought involving food, water, food safety, and agricultural products for human consumption or for use as animal feed, as of the writing of this book. Significant about these cases is the subject matter, food and agricultural products, which are critical pressure points of international trade. The United States, the EU, and Canada have brought, by far, the most cases before the WTO for dispute resolution. Table 6.1 and Figures 6.1 through 6.4 illustrate that the countries bringing WTO cases for dispute resolution are the major industrial nations against, primarily, developing countries.

The trends of the case frequencies, illustrated in Table 6.1 and Figure 6.1, brought by the major industrial countries against less-affluent countries, hold true for all currently pending cases before the WTO. At any given time, the WTO website provides visual illustrations on how many disputes any member country currently participates in, as complainant, respondent, or both. For example, as of October 15, 2013, the United States and the EU have brought complaints against several less-affluent countries, while Brazil and India, for example, are respondents in a multitude of cases brought by the United States and the EU. By comparison, an examination of the same data on March 15, 2017, reveals a similar picture. The arrows point largely in the same directions and the red lines, symbolizing complaints, versus the blue ones, symbolizing responses, are virtually unchanged.

## 6.2    Engineering dispute resolution for the proliferation of GMOs

### 6.2.1    Treaties and challenges

This section focuses on the treaties and challenges relevant to the trivialization through proliferation of GMOs, thereby providing a brief survey of the most relevant treaties and their scope. Although the WTO is responsible for the administration of numerous treaties (see Figure 6.5), not all of them are typically invoked in food and agriculture disputes (see Tables 6.2 and 6.3). By its own account, "WTO members have taken steps to reform the agriculture sector and to address the subsidies and high trade barriers that distort agricultural trade. The overall aim is to establish a fairer trading system that will increase market access and improve the livelihoods of farmers around the world."[7] By creating an enabling

---

[7] WTO, Agriculture, https://www.wto.org/english/tratop_e/agric_e/agric_e.htm (last accessed March 15, 2017).

***Table 6.1*** Dispute resolution case frequencies about food by countries, in tabular form

| | Dispute initiator | Dispute respondent | Total cases by selected |
|---|---|---|---|
| United States | 40 | 75 | 115 |
| European | 23 | 89 | 112 |
| Canada | 16 | 43 | 59 |
| Australia | 5 | 44 | 49 |
| Japan | 0 | 46 | 46 |
| China | 2 | 41 | 43 |
| Chile | 7 | 33 | 40 |
| Mexico | 6 | 32 | 38 |
| Brazil | 12 | 25 | 37 |
| Thailand | 7 | 28 | 35 |
| Argentina | 14 | 19 | 33 |
| India | 3 | 28 | 31 |
| New Zealand | 4 | 25 | 29 |
| Korea, Republic of | 1 | 24 | 25 |
| Colombia | 3 | 16 | 19 |
| Guatemala | 5 | 10 | 15 |
| Norway | 1 | 14 | 15 |
| Peru | 3 | 11 | 14 |
| Ecuador | 3 | 9 | 12 |
| Philippines | 3 | 7 | 10 |
| Venezuela | 0 | 9 | 9 |
| Uruguay | 2 | 6 | 8 |
| Costa Rica | 3 | 4 | 7 |
| Nicaragua | 1 | 6 | 7 |
| Viet Nam | 2 | 4 | 6 |
| Trinidad and Tobago | 0 | 6 | 6 |
| Panama | 2 | 2 | 4 |
| Sti Lanka | 1 | 2 | 3 |

The column in blue shows how often the country was the dispute initiator; the column in green, how often the country was the dispute respondent; and the column in white shows the total number of cases brought by that country. Data last updated in October 2013.

*Figure 6.1* Map of disputes between WTO members on October 15, 2013, with the United States as complainants. The red arrows indicate against which other countries, the United States or the European Union, respectively, brought cases, and the numbers reflect the numbers of currently pending cases. These maps are not limited by subjects or agreements and merely represent the number of presently pending cases. (From WTO, Map of disputes between WTO Members, © World Trade Organization (WTO) 2017, https://www.wto.org/english/tratop_e/dispu_e/dispu_maps_e.htm.)

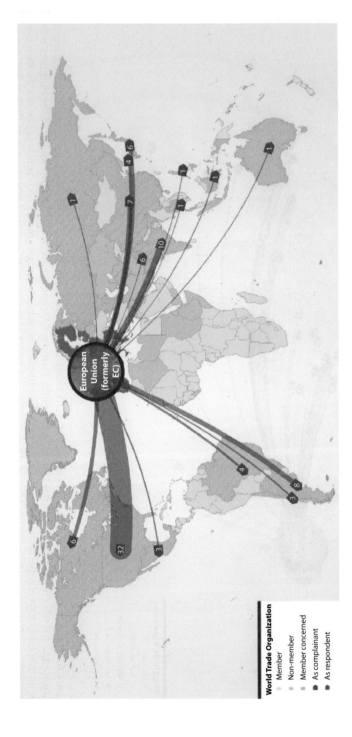

*Figure 6.2* Map of disputes between WTO members on October 15, 2013, with European Union as complainant. (From WTO, Map of disputes between WTO Members, © World Trade Organization (WTO) 2017, https://www.wto.org/english/tratop_e/dispu_e/dispu_maps_e.htm.)

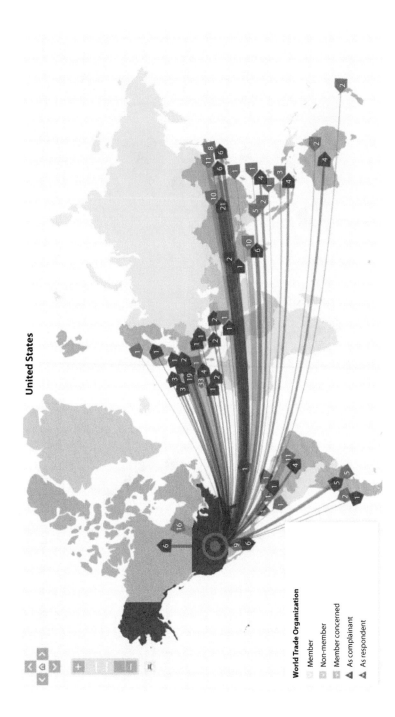

*Figure 6.3* Map of disputes between WTO members on March 15, 2017, with the United States as complainant and respondent. (From WTO, Map of disputes between WTO Members, © World Trade Organization (WTO) 2017, https://www.wto.org/english/tratop_e/dispu_e/dispu_maps_e.htm.)

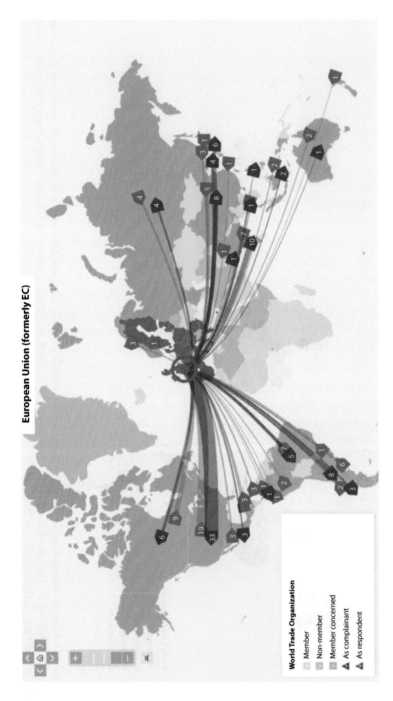

*Figure 6.4* Map of disputes between WTO members on March 15, 2017, with European Union as complainant (red) and respondent (blue). (From WTO, Map of disputes between WTO Members, © World Trade Organization (WTO) 2017, https://www.wto.org/english/tratop_e/dispu_e/dispu_maps_e.htm.)

environment for the trivialization of GMOs and their proliferation in a manner congruent with Figure 2.1, the WTO has not always been success-ful in living up to its mission. Admittedly, agriculture has undergone the so-called green revolution, a reform centering on the industrialization of food production largely benefiting from far-reaching international trade in commodity crops. The true distortion of agricultural trade, however, is misguided and misunderstood. In the context of the WTO's efforts noted earlier, anything that stands in the way of trade allegedly distorts it. While this may be so from an economic standpoint, the shift in perspec-tive here requires a critical evaluation of agricultural trade to meet its ultimate goal of feeding the world—which increased trade counteracts rather than facilitates.[8]

The WTO Agreement on Agriculture, ratified in 1995, "represents a significant step towards reforming agricultural trade and making it fairer and more competitive."[9] According to the WTO, it covers,

1. *Market access*: The use of trade restrictions, such as tariffs on imports.
2. *Domestic support*: The use of subsidies and other support programs that directly stimulate production and distort trade.
3. *Export competition*: The use of export subsidies and other govern-ment support programs that subsidize exports.[10]

In addition, "[u]nder the Agreement, WTO members agree to 'sched-ules' or lists of commitments that set limits on the tariffs they can apply to individual products and on levels of domestic support and export subsidies."[11] Thus, the Agreement on Agriculture is the quintessential international tool to promote the proliferation of GMOs in the name of smooth trade. On the flip side, it is the globally applicable tool to add food miles to commodity crops that are part of the industrialized GMO agri-culture covered in Chapter 1.

The other relevant treaty for food and agriculture is one "on how governments can apply food safety and animal and plant health mea-sures (sanitary and phytosanitary or SPS measures)," the Sanitary and Phytosanitary (SPS) Agreement.[12] Seemingly focused on food safety,

---

[8] Gabriela Steier, A window of opportunity for GMO regulation: Achieving food integrity through cap-and-trade models from climate policy for GMO regulation, 34 *Pace Envtl L. Rev.* (forthcoming July 2017).

[9] WTO, Agriculture (undated), https://www.wto.org/english/tratop_e/agric_e/agric_e. htm (last accessed Mar. 15, 2017).

[10] 3. Id.

[11] Id.

[12] WTO, Sanitary and phytosanitary measures (undated), https://www.wto.org/english/ tratop_e/sps_e/sps_e.htm (last accessed Mar. 15, 2017).

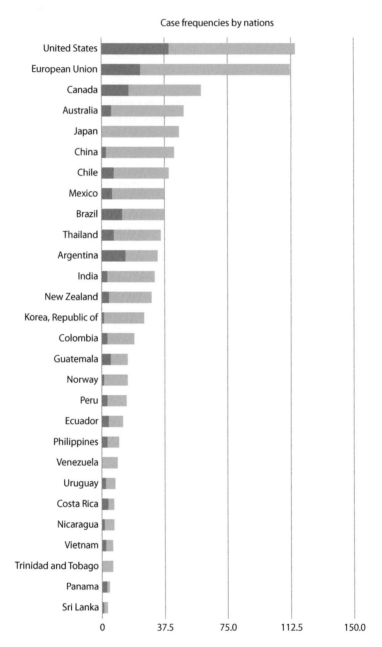

*Figure 6.5* WTO dispute resolution case frequencies about food by countries illustrated in a bar chart. The column in blue shows how often the country was the dispute initiator; the column in green, how often the country was the dispute respondent; and the overall length of the bars show the total number of cases brought by that country.

*Table 6.2* Most frequently cited agreements in WTO disputes on issues of food, water, agricultural products, and food safety (last updated October 2013)

| Agreement | Frequency |
| --- | --- |
| GATT Art. III | 70 |
| GATT Art. I | 59 |
| GATT Art. II | 47 |
| Agriculture Art. 4 | 41 |
| Anti-dumping (Art. VI GATT 1994) | 40 |
| SPS | 37 |
| TBT | 29 |
| Import licensing | 23 |
| Safeguards (Art. 2 and 3, 4, or 5) | 18 |
| Customs valuation (Art. VII GATT 1994) | 13 |
| Agriculture Art. 3 | 12 |
| Agreement establishing the WTO | 10 |

the SPS Agreement "[e]nsur[es] safe trading without unnecessary restrictions."[13] Despite being beyond the scope of this book to evaluate what "unnecessary restrictions" are, the clashes between the EU and the United States, as well as numerous other countries including Brazil, Argentina, Canada, and India, show that the SPS Agreement has a complex and complicated set of values under its purview. With agricultural

*Table 6.3*   Most frequently cited agreements in WTO disputes on issues of food, water, agricultural products, and food safety (last updated October 2013)

| Agreement | Frequency |
| --- | --- |
| GATT | 181 |
| Agriculture Art. 1–19 | 57 |
| SPS | 37 |
| TBT | 29 |
| Subsidies and countervailing measures | 22 |
| Agreement establishing the WTO | 10 |
| TRIMs | 8 |
| Services (GATS) | 4 |
| TRIPS | 2 |

---

[13] WTO, Sanitary and phytosanitary measures: Ensuring safe trading without unnecessary restrictions (undated), https://www.wto.org/english/thewto_e/20y_e/sps_brochure20y_e.pdf (last accessed Mar. 15, 2017).

products worth over US$1765 billion traded in 2013[14] alone, the importance of the SPS Agreement must not be understated. However, zooming back in on the question of trivializing the dangers of GMOs by facilitating their proliferation, the purpose of the SPS Agreement is suspiciously similar to the industry-friendly Food and Drug Administration (FDA) guidance, described in Chapter 5, which discounts consumer protection at the expense of free trade prioritization.

Specifically, with the goal to "achieve a balance between the right of WTO members to implement *legitimate health protection* policies and the goal of allowing the *smooth flow of goods* across international borders without *unnecessary* restrictions,"[15] (emphasis added) one can easily envision how industry lobbyists construe "legitimate," "smooth," and "unnecessary" in this context. These terms create loopholes and opportunities to whitewash a neoliberal and capitalist food system aimed at boosting the bottom line of BigAg rather than nurturing the people. Thus, the SPS Agreement is another tool in the arsenal that fires up the cycle depicted in Figure 2.1, promoting the proliferation of GMOs with little concern for food security, sovereignty, and true consumer-centered food safety.

Alongside the Agriculture and SPS Agreements is the Technical Barriers to Trade (TBT) Agreement, which "aims to ensure that technical regulations, standards, and conformity assessment procedures are non-discriminatory and do not create unnecessary obstacles to trade."[16] Similar to the SPS Agreement, "it recognises WTO members' right to implement measures to achieve legitimate policy objectives, such as the protection of human health and safety, or protection of the environment" but, again dilutes these rights by "strongly encourag[ing] members to base their measures on international standards as a means to facilitate trade."[17] Even worse, instead of gearing transparency toward consumers and democratic checks of its work, the TBT Agreement, "[t]hrough its transparency provisions, ...also aims to create a predictable trading environment."[18]

Another treaty is the General Agreement on Tariffs and Trade (GATT) of 1994, which basically created the WTO with the purpose to achieve "substantial reduction of tariffs and other trade barriers and the elimination of preferences, on a reciprocal and mutually advantageous basis."[19] It is often cited in food- and agriculture-related cases, but relevant in principle.

---

[14] Id.

[15] Id.

[16] WTO, Technical barriers to trade (undated), https://www.wto.org/english/tratop_e/tbt_e/tbt_e.htm (last accessed Mar. 15, 2017).

[17] Id.

[18] Id.

[19] GATT/WTO, *Duke Law Guide* (undated), https://law.duke.edu/lib/researchguides/gatt/ (last accessed Mar. 15, 2017).

## 6.2.2 The biotech cases: Storming the EU by GMOs

The aforementioned treaties are herein criticized for being tools used to promote the proliferation of GMOs by trivializing their risks with vague language. In this section, a series of cases illustrates this contention. Specifically, these cases are selected from over 600 (see the appendix to this chapter) as ones that bulldozed restrictions to the proliferation of GMOs by trivializing their dangers in the name of free trade.

In 2003, challenges by the United States, Canada, and Argentina against the EU were consolidated in three cases: Disputes DS291, DS292, and DS293.[20] This series of three high-stake cases was brought by the United States, Canada, and Argentina against the EU with an array of named third parties requesting that the then-existing moratorium on biotech product import to the EU be lifted. Although the EU issued a ban on the approval of biotech products from the United States, the complainants' victory at the WTO dispute resolution process resulted in import permits of GMOs into the EU. In response to the complaint that the moratorium, which the EU denied before the panel, has restricted imports of agricultural and food products from the United States, the EU argued that the measures to preserve biodiversity are the competence of the Cartagena Protocol on Biosafety, following the risk management analysis "harmful until proven safe" under the Cartagena Protocol to avoid the import of GMO crops to the EU. As noted in Section 5.2, the Cartagena Protocol implements the precautionary approach to GMO approval, which explains why there was a moratorium in the first place.

Despite the fact that none of the complainants are signatories to the Cartagena Protocol on Biosafety, the WTO dispute resolution panel found that the EU acted inconsistently with its obligations under the SPS Agreement by causing undue delays in the completion of the approval procedures for the biotech products and with regard to all of the safeguard measures, which the panel found were not supported by sufficient scientific evidence based on risk assessments satisfying the definition of the SPS Agreement. Here, the WTO panel was not swayed by the precautionary approach and trivialized the risks of GMOs, allowing them to be introduced in the EU to ensure smooth trade flows.

A few years later, the same treaties were invoked with a similar effect in what has come to be known as the *Hormone-Treated Meat Import from the U.S. to the European Communities Challenges by the U.S. against the European Communities in Dispute DS26.*[21]

---

[20] WTO, European communities—Measures affecting the approval and marketing of biotech products (2010), https://www.wto.org/english/tratop_e/dispu_e/cases_e/ds291_e. htm.

[21] WTO, DS 26—European Communities—Measures Concerning Meat and Meat Products (Hormones), https://www.wto.org/english/tratop_e/dispu_e/cases_e/ds26_e.htm.

In 2009, the United States filed this complaint against the EU under the Agriculture Act, GATT, and the SPS and TBT Agreements, claiming that the European Council Directives aimed at banning and restricting the use and import of US-American meat and meat products from animals treated with hormone-based veterinary drugs. The other major meat producers worldwide, Australia, Canada, New Zealand and Norway, joined preserving their rights as third parties.

The directives cited in the complaint concerned three naturally occurring and three artificial hormones that were administered to livestock for therapeutic use, but, more importantly, growth promotion. European consumers became increasingly concerned about the safety to human health from the consumption of hormone-treated meat, which could lead to hormonal irregularities and tumorigenesis. The panel's analysis of the Codex Alimentarius standards, the agreements cited, and the allowable daily intake of hormones through meat consumption concluded with the panel's recommendation that the European Communities bring its measures in dispute into conformity with its obligations under the SPS Agreement, thereby allowing for the import of hormone-treated meat from the United States. Again, the WTO's panel bulldozed the EU's resistance to hormone-treated meat. Although this is not an example of GMOs, it is a fitting analogy to illustrate how the treaties are being used and how the criticisms issued in Chapter 5 were acted out in these cases.

By way of a final example, again not necessarily one involving GMOs but, nonetheless, instructive, Dispute DS308 titled *Mexican Import Taxes on Soft-Drinks from the US: Complaint by the US against Mexico*[22] provides further support for the contentions of this chapter. In 2004, the United States filed a complaint under GATT against Mexico's import taxes on US soft drinks imports that used any sweetener other than cane sugar. The taxes, in the form of import and distribution taxes, were about 20%. Canada, China, the European Communities, Guatemala, and Japan reserved their third-party rights. Mexico argued that the soft-drink tax was "necessary to secure compliance" by the United States with its obligations under the North American Free Trade Agreement (NAFTA), an international agreement that is a law consistent with the provisions of the GATT. However, the United States argued that the soft-drink tax measure was not intended to afford protection to domestic production within the meaning of Article III of the GATT. The WTO panel ultimately recommended that Mexico withdraw the tax, and Mexico complied.

This case shows how the United States managed to bully Mexico into opening its market to US-American soft-drink companies. Yet another example that illustrates what Sections 6.1 and 6.2 describe brings the

[22] WTO, Dispute DS308 Mexico—Tax Measures on Soft Drinks and Other Beverages, https://www.wto.org/english/tratop_e/dispu_e/cases_e/ds308_e.htm.

discussion full circle, alluding again to Figure 2.1. Instead of protecting food and meeting its policy mandates, the WTO merely provides mechanisms to "get big or get out," virtually escalating agricultural exceptionalism from the United States to the global market.

## 6.3   Summary

This chapter explored what happens when countries disagree over the risks of GMOs and the market shifts resulting from GMO trade. Disputes over such trade usually go before the WTO Dispute Resolution Body, which has examined a total of 623 cases involving food, water, food safety, and agricultural products for human consumption or for use as animal feed, as of the writing of this book. What is significant about these cases is the subject matter, food and agricultural products, which are critical pressure points of international trade. Moreover, this chapter examined the treaties and challenges relevant to the trivialization through proliferation of GMOs, thereby providing a brief survey of the most relevant treaties and their scope, such as the Agreement on Trade-Related Aspects of Intellectual Property Rights (TRIPS), the Agreement on the Application of Sanitary and Phytosanitary Measures (SPS), and the Agreement on Technical Barriers to Trade (TBT).

## Appendix: The disputes

WTO dispute digest on issues of food, agricultural products, food safety, and public health.

| Resolution year | Case | Dispute date and status | Parties | | Agreements cited | Summary |
| | | | Complainant | Respondent | | |
| --- | --- | --- | --- | --- | --- | --- |
| 1996 | **DISPUTE DS41** Korea—Measures Concerning Inspection of Agricultural Products | In consultations on 24 May 1996 (up-to-date at 24 February 2010) | United States | Korea, Republic of | Agriculture: Art. 4 GATT 1994: Art. III, XI Sanitary and Phytosanitary Measures (SPS): Art. 2, 5, 8 Technical Barriers to Trade (TBT): Art. 2, 5, 6 | On 24 May 1996, the United States requested consultations with Korea concerning testing, inspection, and other measures required for the importation of agricultural products into Korea. The United States claimed that these measures restrict imports and appear to be inconsistent with the WTO Agreement. Violations of GATT Articles III and XI, SPS Articles 2, 5 and 8, TBT Articles 2, 5 and 6, and Article 4 of the Agreement on Agriculture are alleged. The United States requested consultations with Korea on similar issues on 4 April 1995 (WT/DS3). |

*(Continued)*

| Resolution year | Case | Dispute date and status | Parties | | Agreements cited | Summary |
|---|---|---|---|---|---|---|
| | | | Complainant | Respondent | | |
| 1995 | **DISPUTE DS3** Korea—Measures Concerning the Testing and Inspection of Agricultural Products | In consultations on 4 April 1995 (up-to-date at 24 February 2010) | United States | Korea, Republic of | Agriculture: Art. 4 GATT 1994: Art. III, XI Sanitary and Phytosanitary Measures (SPS): Art. 2, 5 Technical Barriers to Trade (TBT): Art. 5, 6 | Complaint by the United States. On 6 April 1995, the United States requested consultations with Korea involving testing and inspection requirements with respect to imports of agricultural products into Korea. The measures are alleged to be in violation of GATT Articles III or XI, Articles 2 and 5 of the Agreement on Sanitary and Phytosanitary Measures (SPS), TBT Articles 5 and 6 and Agriculture Article 4. (See WT/DS41). |

*(Continued)*

| Resolution year | Case | Dispute date and status | Parties | | Agreements cited | Summary |
|---|---|---|---|---|---|---|
| | | | Complainant | Respondent | | |
| 1996 | **DISPUTE DS35** Hungary—Export Subsidies in respect of Agricultural Products | Settled or terminated (withdrawn, mutually agreed solution) on 30 July 1997 (up-to-date at 24 February 2010) | Argentina; Australia; Canada; New Zealand; Thailand; United States | Hungary **Third parties:** Canada; Japan; Thailand; Uruguay | Agriculture: Art. 3.3, Part V | This request, dated 27 March 1996, claims that Hungary violated the Agreement on Agriculture (Article 3.3 and Part V) by providing export subsidies in respect of agricultural products not specified in its Schedule, as well as by providing agricultural export subsidies in excess of its commitment levels. **Panel and appellate body proceedings** On 9 January 1997, Argentina, Australia, New Zealand, and the United States requested the establishment of a panel. At its meeting on 25 February 1997, the DSB established a panel. Canada, Japan, Thailand, and Uruguay reserved their third-party rights to the dispute. **Mutually agreed solution** At the DSB meeting on 30 July 1997, Australia, on behalf of all the complainants, notified the DSB that the parties to the dispute had reached a mutually agreed solution, which required Hungary to seek a waiver of certain of its WTO obligations. Pending adoption of the waiver, the complaint was not formally withdrawn. |

*(Continued)*

| Resolution year | Case | Dispute date and status | Parties Complainant | Parties Respondent | Agreements cited | Summary |
|---|---|---|---|---|---|---|
| 2001 | **DISPUTE DS76** Japan—Measures Affecting Agricultural Products | Mutually acceptable solution on implementation notified on 25 September 2001 (up-to-date at 24 February 2010) | United States | Japan **Third parties:** Brazil; European Communities; Hungary | Agriculture: Art. 4 GATT 1994: Art. XI Sanitary and Phytosanitary Measures (SPS): Art. 2, 4, 5, 7, 8 | **Measure at issue:** Varietal testing requirement (Japan's Plant Protection Law), under which the import of certain plants was prohibited because of the possibility of their becoming potential hosts of codling moth. On 7 April 1997, the United States requested consultations with Japan in respect of the latter's prohibition, under quarantine measures, of imports of certain agricultural products. The US alleged that Japan prohibits the importation of each variety of a product requiring quarantine treatment until the quarantine treatment has been tested for that variety, even if the treatment has proved to be effective for other varieties of the same product. The United States alleged violations of Articles 2, 5, and 8 of the SPS Agreement, Article XI of GATT 1994, and Article 4 of the Agreement on Agriculture. In addition, the United States made a claim for nullification and impairment of benefits. See http://www.wto.org/english/tratop_e/dispu_e/cases_e/1pagesum_e/ds76sum_e.pdf. |

*(Continued)*

| Resolution year | Case | Dispute date and status | Parties | | Agreements cited | Summary |
| | | | Complainant | Respondent | | |
| --- | --- | --- | --- | --- | --- | --- |
| 2006 | DISPUTE DS174 European Communities—Protection of Trademarks and Geographical Indications for Agricultural Products and Foodstuffs | Implementation notified by respondent on 21 April 2006 (up-to-date at 24 February 2010) | United States | EU **Third parties:** Argentina; Australia; Brazil; Canada; China; Chinese Taipei; Colombia; Guatemala; India; Mexico; New Zealand; Turkey | GATT 1994: Art. I, III:4 Intellectual Property (TRIPS): Art. 1.1, 2, 2.1, 3, 3.1, 4, 16, 16.1, 20, 22, 22.1, 22.2, 24, 24.5, 41.1, 41.2, 41.4, 42, 44.1, 63, 63.1, 63.3, 65, 65.1 | **Measure at issue:** EC Regulation related to the protection of geographical indications and designations of origin ("GIs"). **Product at issue:** Agricultural products and foodstuffs affected by the EC Regulation. **Availability of protection:** The Panel found that the equivalence and reciprocity conditions in respect of GI protection under the EC Regulation 3 violated the national treatment obligations under TRIPS Art. 3.1 and GATT Art. III:4 by according less favorable treatment to non–EC nationals and products, than to EC nationals and products. By providing, "formally identical," but in fact different procedures based on the location of a GI, the European Communities in fact modified the "effective equality of opportunities" between different nationals and products to the detriment of non–EC nationals and products. |

*(Continued)*

| Resolution year | Case | Dispute date and status | Parties | | Agreements cited | Summary |
|---|---|---|---|---|---|---|
| | | | Complainant | Respondent | | |
| | | | | | | **Application and objection procedures:** The Panel found that the Regulation's procedures requiring non-EC nationals, or persons resident or established in non-EC countries, to file an application or objection in the European Communities through their own government (but not directly with EC member states) provided formally less favorable treatment to other nationals and products in violation of TRIPS Art. 3.1 and GATT Art. III:4, and that the GATT violation was not justified by Art. XX(d). |

*(Continued)*

| Resolution year | Case | Dispute date and status | Parties | | Agreements cited | Summary |
| --- | --- | --- | --- | --- | --- | --- |
| | | | Complainant | Respondent | | |
| | | | | | | **Inspection structures:** In the US Report, the Panel found that the Regulation's requirement that third-country governments provide a declaration that structures to inspect compliance with GI registration were established on its territory violated TRIPS Art. 3.1 and GATT Art. III:4 by providing an "extra hurdle" to applicants for GIs in third countries and their products, and that the GATT violation was not justified by Art. XX(d). In the Australian report, the Panel found that these inspection structures did not constitute a "technical regulation" within the meaning of the TBT Agreement. **Relationship between GIs and (prior) trademarks:** |

*(Continued)*

| Resolution year | Case | Dispute date and status | Parties | | Agreements cited | Summary |
|---|---|---|---|---|---|---|
| | | | Complainant | Respondent | | |
| | | | | | | TRIPS Arts. 16.1, 17, 24.3 and 24.5: The Panel found that Art. 16.1 obliges Members to make available to trademark owners a right against certain uses, including uses as a GI. Art. 24.5 provided no defence, as it creates an exception to GI rights, not trademark rights. Art. 24.3 only grandfathers individual GIs, not systems of GI protection. Therefore, the Panel initially concluded that the EC Regulation was inconsistent with Art. 16.1 as it limited the availability of trademark rights where the trademark was used as a GI. However, the Panel ultimately found that the Regulation, on the basis of the evidence presented, was justified under Art. 17, which permits Members to provide exceptions to trademark rights. See http://www.wto.org/english/tratop_e/dispu_e/cases_e/1pagesum_e/ds174sum_e.pdf |

(*Continued*)

| Resolution year | Case | Dispute date and status | Parties | | Agreements cited | Summary |
|---|---|---|---|---|---|---|
| | | | Complainant | Respondent | | |
| 2007 | DISPUTE DS207 Chile—Price Band System and Safeguard Measures Relating to Certain Agricultural Products | Compliance proceedings completed with finding(s) of non-compliance on 22 May 2007 (up-to-date at 24 February 2010) | Argentina | Chile **Third parties:** Australia; Brazil; Canada; China; Colombia; Costa Rica; European Communities; Ecuador; El Salvador; Guatemala; Honduras; Japan; Nicaragua; Paraguay; Peru; Thailand; Venezuela, Bolivarian Republic of; United States | Agriculture: Art. 4 GATT 1994: Art. II, XIX:1 Safeguards: Art. 2, 3, 4, 5, 6, 12 | **Measure at issue:** Chile's Price Band System, governed by Rules on the Importation of Goods, through which the tariff rate for products at issue could be adjusted to international price developments if the price fell below a lower price band or rose beyond an upper price band. **Product at issue:** Wheat, wheat flour, sugar, and edible vegetable oils from Argentina **DSU Art. 11 (standard of review):** The Appellate Body reversed the Panel's findings under GATT Art. II:1(b), second sentence, on the grounds that it was a claim that had not been raised by Argentina in its panel request or any subsequent submissions, and the Panel, by assessing a provision that was not part of the matter before it, acted ultra petita and in violation of DSU Art. 11. The Appellate Body also stated that consideration by a Panel of claims not raised by the complainant deprived Chile of its due process rights under the DSU. |

*(Continued)*

| Resolution year | Case | Dispute date and status | Parties | | Agreements cited | Summary |
|---|---|---|---|---|---|---|
| | | | Complainant | Respondent | | |
| | | | | | | AA Art. 4.2, footnote 1 (market access): The Appellate Body reversed the Panel's findings that the term "ordinary customs duty" was to be understood as referring to "a customs duty which is not applied to factors of an exogenous nature" and Chile's price brand system was not an "ordinary customs duty," as it was assessed on the basis of exogenous price factors. The Appellate Body, however, upheld the Panel's finding that Chile's price band system was designed and operated as a border measure sufficiently similar to "variable import levies" and "minimum import prices" within the meaning of footnote 1 and therefore prohibited by Art. 4.2. Thus, the Appellate Body concluded that Chile's price band system was inconsistent with Art. 4.2. See http://www.wto.org/ english/tratop_e/dispu_e/ cases_e/1pagesum_e/ ds207sum_e.pdf |

*(Continued)*

| Resolution year | Case | Dispute date and status | Parties Complainant | Parties Respondent | Agreements cited | Summary |
|---|---|---|---|---|---|---|
| 2001 | DISPUTE DS220 Chile—Price Band System and Safeguard Measures Relating to Certain Agricultural Products | In consultations on 5 January 2001 (up-to-date at 24 February 2010) | Guatemala | Chile | Agriculture: Art. 4 GATT 1994: Art. II, XIX Safeguards: Art. 2, 3, 4, 5, 6, 8, 12 | On 5 January 2001, Guatemala requested consultations with Chile concerning: 1. The Chilean legislation regarding safeguards and price band systems, including Law 18.525, as subsequently amended by Law 18.591 and Law 19.546, as well as implementing regulations and complementary and/or amending provisions. 2. The initiation of an investigation regarding products subject to the price band system contained in notification G/SG/N/6/CHL/2; the conduct of the investigation, the preliminary determination contained in notification G/SG/N/7/CHL/2/Suppl.1, and the definitive determination contained in notifications G/SG/N/8/CHL/1, G/SG/N/10/CHL/1, G/SG/N/8/CHL/1/Suppl.1 and G/SG/N/10/CHL/1/Suppl.1; these notifications indicate that wheat, wheat flour, sugar and edible vegetable oils are subject to said safeguard measures. |

*(Continued)*

| Resolution year | Case | Dispute date and status | Parties | | Agreements cited | Summary |
|---|---|---|---|---|---|---|
| | | | Complainant | Respondent | | |
| | | | | | | 3. The request for an extension of these measures contained in notifications G/SG/N/10/CHL/1/Suppl.2 and G/SG/N/10/CHL/1/Suppl.2/Corr.1. |
| | | | | | | 4. Guatemala considered that the measures referred to: |
| | | | | | | 5. Under (1) are inconsistent with, *inter alia*, Article II of GATT 1994 and Article 4 of the Safeguards Agreement. |
| | | | | | | 6. Under (2) are inconsistent with, *inter alia*, Articles 2, 3, 4, 5, 6 and 12 of the Safeguards Agreement, and Article XIX:1 of GATT 1994. |
| | | | | | | 7. Under (3) appears to be inconsistent with, *inter alia*, Chile's obligations under GATT 1994 and Articles 2, 3, 4, 5, 6, 8 and 12 of the Safeguards Agreement. |

*(Continued)*

| Resolution year | Case | Dispute date and status | Parties | | Agreements cited | Summary |
|---|---|---|---|---|---|---|
| | | | Complainant | Respondent | | |
| 2005 | DISPUTE DS245 Japan—Measures Affecting the Importation of Apples | Mutually acceptable solution on implementation notified on 30 August 2005 (up-to-date at 24 February 2010) | United States | Japan **Third parties:** Australia; Brazil; China; Chinese Taipei; European Communities; New Zealand | Agriculture: Art. 4.2, 14 GATT 1994: Art. XI Sanitary and Phytosanitary Measures (SPS): Art. 2.2, 2.3, 5.1, 5.2, 5.3, 5.5, 5.6, 6.1, 6.2, 7, Annex B | **Measure at issue:** Certain Japanese measures restricting imports of apples on the basis of concerns about the risk of transmission of fire blight bacterium. **Product at issue:** Apples from the United States SPS Art. 2.2 (sufficient scientific evidence): The Appellate Body upheld the Panel's finding that the measure was maintained "without sufficient scientific evidence" inconsistently with Art. 2.2, as there was a clear disproportion (and thus no rational or objective relationship) between Japan's measure and the "negligible risk" identified on the basis of the scientific evidence. |

*(Continued)*

| Resolution year | Case | Dispute date and status | Parties | | Agreements cited | Summary |
|---|---|---|---|---|---|---|
| | | | Complainant | Respondent | | |

| | |
|---|---|
| Parties | |
| Complainant | Respondent |

**SPS Art. 5.7 (provisional measure):** The Appellate Body upheld the Panel's finding that the measure was not a provisional measure justified within the meaning of Art. 5.7, as the measure was not imposed in respect of a situation "where relevant scientific evidence is insufficient." Having noted that the pertinent question under Art. 5.7 is whether the body of available scientific evidence does not allow, in quantitative or qualitative terms, the performance of an adequate assessment of risks, as required under Art. 5.1 and as defined in Annex A of the SPS Agreement, the Appellate Body found that in light of the Panel's finding of a large quantity of high-quality scientific evidence describing the risk of transmission of fire blight through apple fruit, there was "the body of available scientific evidence" in this case that would allow "the evaluation of the likelihood of entry, establishment or spread" of fire blight in Japan through apples exported from the United States.

(*Continued*)

| Resolution year | Case | Dispute date and status | Parties | | Agreements cited | Summary |
|---|---|---|---|---|---|---|
| | | | Complainant | Respondent | | |
| | | | | | | **SPS Art. 5.1 (risk assessment):** The Appellate Body upheld the Panel's finding that the measure was not based on a risk assessment as required under Art. 5.1 because the pest risk analysis relied on by Japan (i.e., "1999 PRA") failed to evaluate (1) the likelihood of entry, establishment or spread of fire blight specifically through apple fruit; and (2) the likelihood of entry "according to the SPS measures that might be applied." In this regard, the Appellate Body noted that the obligation to conduct an assessment of "risk" under Art. 5.1 is not satisfied merely by a general discussion of the disease sought to be avoided by the imposition of the SPS measure, rather an evaluation of the risk must connect the possibility of adverse effects with an antecedent or cause (i.e., in this case, transmission of fire blight "through apple fruit"). Also, the Appellate Body upheld the Panel's view that the definition of "risk assessment" requires that the evaluation of the entry, establishment or spread of a disease be conducted according to the sanitary or phytosanitary measures which might be applied, not merely measures which are being currently applied. |

*(Continued)*

| Resolution year | Case | Dispute date and status | Parties | | Agreements cited | Summary |
|---|---|---|---|---|---|---|
| | | | Complainant | Respondent | | |
| | | | | | | See http://www.wto.org/english/tratop_e/dispu_e/cases_e/1pagesum_e/ds245sum_e.pdf |
| 2004 | DISPUTE DS250 United States—Equalizing Excise Tax Imposed by Florida on Processed Orange and Grapefruit Products | Settled or terminated (withdrawn, mutually agreed solution) on 28 May 2004 (up-to-date at 24 February 2010) | Brazil | United States **Third parties:** Chile; European Communities; Mexico; Paraguay | GATT 1994: Art. II, III | On 20 March 2002, Brazil requested consultations with the United States concerning the so-called "Equalizing Excise Tax" imposed by the State of Florida on processed orange and grapefruit products produced from citrus fruit grown outside the United States (Section 601.155 Florida Statutes). Brazil indicated that since 1970, the state of Florida had imposed, pursuant to Section 601.155 of the Florida Statutes, an "equalizing excise tax" on processed orange and processed grapefruit products, in amounts determined by the Florida Department of Citrus. However, the statute by its terms—Section 601.155(5), Florida Statutes—exempted from the tax products "produced in whole or in part from citrus fruit grown within the United States." In the view of Brazil the incidence of this tax on imported processed citrus products and not on domestic products on its face constituted a violation of Articles II:1(a), III:1 and III:2 of GATT 1994.

*(Continued)* |

| Resolution year | Case | Dispute date and status | Parties | | Agreements cited | Summary |
|---|---|---|---|---|---|---|
| | | | Complainant | Respondent | | |
| | | | | | | Brazil contended that the impact of the Florida equalizing excise tax had been to provide protection and support to domestic-processed citrus products and to restrain the importation of processed citrus products into Florida. Since processed citrus products, principally in the form of frozen concentrated orange juice, were among Brazil's most significant exports to the United States, Brazil was of the view that the restraint on their importation by the State of Florida constituted a nullification and impairment of benefits accruing to Brazil under GATT 1994. Brazil reserved the right to raise additional factual or legal points related to the aforementioned measure during the course of consultations. On 16 August 2002, Brazil requested the establishment of the panel. At its meeting on 30 August 2002, the DSB deferred the establishment of a panel. *(Continued)* |

| Resolution year | Case | Dispute date and status | Parties | | Agreements cited | Summary |
| | | | Complainant | Respondent | | |
| --- | --- | --- | --- | --- | --- | --- |
| 2002 | DISPUTE DS255 Peru—Tax Treatment on Certain Imported Products | Settled or terminated (withdrawn, mutually agreed solution) on 25 September 2002 (up-to-date at 24 February 2010) | Chile | Peru | GATT 1994: Art. III | On 22 April 2002, Chile requested consultations with Peru in respect of its tax treatment on imports of fresh fruits, vegetables, fish, milk, tea, and other natural products. In particular, Chile explained that before the adoption of Law 27.614, published on 29 December 2001, both the sale in the Peruvian market and the importation into Peru of the products at issue had been exempt from sales tax. Further to the adoption of Law 27.614, the importation into Peru of the products at issue was no longer exempt from sales tax (18 per cent) while the sale of those products in the Peruvian market was still exempt from sales tax. |

*(Continued)*

| Resolution year | Case | Dispute date and status | Parties | | Agreements cited | Summary |
| --- | --- | --- | --- | --- | --- | --- |
| | | | Complainant | Respondent | | |
| | | | | | | Chile considered that the different tax treatment between domestic and imported products constituted a violation by Peru of its national treatment commitments both at bilateral level, Article 19 of the Economic Complementarity Agreement (ECA 38), and at multilateral level (WTO), Article III of GATT 1994. In this regard, Chile claimed that Law 27.614 by providing that the exemption from sales tax only applied to the sale in Peru and not to the importation, was inconsistent with Article III of the GATT 1994. Chile contended that the above measure prejudiced the competitiveness of Chile's exports to Peru of the said products, in particular although not limited to, apples, table grapes, and peaches. |

*(Continued)*

| Resolution year | Case | Dispute date and status | Parties | | Agreements cited | Summary |
|---|---|---|---|---|---|---|
| | | | Complainant | Respondent | | |
| | | | | | | On the grounds that the products affected by the measure at issue were natural goods and thus perishable, Chile was invoking the urgency consultation procedure under Article 4.8 of the DSU and already announced that it would make use of the expeditious Panel and Appellate Body procedures referred to in Article 4.9 of the DSU. On 8 May 2002, the United States requested to join the consultations. On 13 June 2002, Chile requested the establishment of a panel. At its meeting on 24 June 2002, the DSB deferred the establishment of a panel. On 26 July 2002, Chile requested that its second request for the establishment of a panel be removed from the agenda of the DSB meeting on 29 July 2002. Withdrawal/termination: On 25 September 2002, Chile informed the DSB that it was withdrawing this complaint as Peru had repealed Article 2 of Law 27.614 and as a result the disputed measures had disappeared. |

*(Continued)*

| Resolution year | Case | Dispute date and status | Parties | | Agreements cited | Summary |
|---|---|---|---|---|---|---|
| | | | Complainant | Respondent | | |
| 2005 | DISPUTE DS265 European Communities—Export Subsidies on Sugar | Report(s) adopted, with recommendation to bring measure(s) into conformity on 19 May 2005 (up-to-date at 24 February 2010) | Australia | European Communities **Third parties:** Barbados; Belize; Brazil; Canada; China; Colombia; Cuba; Fiji; Guyana; India; Jamaica; Kenya; Madagascar; Malawi; Mauritius; New Zealand; Paraguay; Saint Kitts and Nevis; Swaziland; Tanzania; Thailand; Trinidad and Tobago; United States; Côte d'Ivoire | Agriculture: Art. 3.3, 8, 9.1, 10.1, 11 GATT 1994: Art. III:4, XVI Subsidies and Countervailing Measures: Art. 3.1(a), 3.2 | Australia contended that the EC provides under the above measures export subsidies in excess of the export subsidy commitments that it has specified in Section II of Part IV of its Schedule of Concessions, in relation to "C sugar" and an amount of 1.6 million tons of sugar per year and possibly also sugar in incorporated products. It further alleges that the EC may also be paying a higher per unit subsidy on incorporated products than on the primary product. In addition, under the EC sugar regime, refiners are paid a subsidy, in the form of the intervention price, for refining EC sugar which is not available to imported sugar, thus affording less favorable treatment to imported products. |

*(Continued)*

| Resolution year | Case | Dispute date and status | Parties | | Agreements cited | Summary |
|---|---|---|---|---|---|---|
| | | | Complainant | Respondent | | |
| | | | | | | According to Thailand: The EC sugar regime accords imported sugar a less favorable treatment than that accorded to domestic sugar and provides for subsidies contingent upon the use of domestic over imported products. The EC sugar regime accords export subsidies above its reduction commitment levels specified in Section II of Part IV of the EC's Schedule to the sugar produced in excess of its production quotas (so-called C sugar). The EC provides export subsidies (known as "export refunds") that cover the difference between the world market price and the high prices in the EC for the products in question, thus enabling those products to be exported. |

*(Continued)*

| Resolution year | Case | Dispute date and status | Parties | | Agreements cited | Summary |
|---|---|---|---|---|---|---|
| | | | Complainant | Respondent | | |
| | | | | | | In light of Article 10.3 of the Agreement on Agriculture, which provides that where a Member exports an agricultural product in quantities that exceed its quantity commitment level, that Member will be treated as if it has granted WTO-inconsistent export subsidies for the excess quantities, unless the Member presents adequate evidence to "establish" the contrary, the Panel reached the conclusion that the European Communities had not demonstrated that the exports of C sugar and "ACP/India equivalent" sugar in excess of its annual commitment levels were not subsidized. |

*(Continued)*

| Resolution year | Case | Dispute date and status | Parties | | Agreements cited | Summary |
|---|---|---|---|---|---|---|
| | | | Complainant | Respondent | | |
| | | | | | | The Panel concluded that the European Communities, through its sugar regime, had acted inconsistently with its obligations under Articles 3.3 and 8 of the Agreement on Agriculture, by providing export subsidies within the meaning of Article 9.1(a) and (c) of the Agreement on Agriculture in excess of the quantity commitment level and the budgetary outlay commitment level specified in Section II, Part IV of Schedule CXL. At its meeting of 13 December 2004, following a request from all the parties, the DSB agreed to extend the 60-day period for the adoption of the Panel report until 31 January 2005. On 13 January 2005, the European Communities notified its intention to appeal certain issues of law and legal interpretations developed by the Panel. On 28 April 2005, the report of the Appellate Body was circulated. The Appellate Body found that: |

*(Continued)*

| Resolution year | Case | Dispute date and status | Parties | | Agreements cited | Summary |
| --- | --- | --- | --- | --- | --- | --- |
| | | | Complainant | Respondent | | |
| | | | | | | • Footnote 1 does not enlarge or otherwise modify the European Communities' commitment levels as specified in its Schedule; Footnote 1 does not contain a commitment to limit subsidization of exports of ACP/India equivalent sugar; and that Footnote 1 is inconsistent with the Agreement on Agriculture, because it does not contain a budgetary outlay commitment and does not subject subsidized exports of ACP/India equivalent sugar to reduction commitments. |

*(Continued)*

| Resolution year | Case | Dispute date and status | Parties | | Agreements cited | Summary |
|---|---|---|---|---|---|---|
| | | | Complainant | Respondent | | |
| | | | | | | • In the particular circumstances of this dispute, there is a "payment" in the form of a transfer of financial resources from the high revenues resulting from sales of A and B sugar, to the export production of C sugar, within the meaning of Article 9.1(c) of the Agreement on Agriculture; such payments were "on the export" within the meaning of Article 9.1(c), because C sugar, under European Communities' law, must be exported, and that the European Communities had acted inconsistently with Articles 3.3 and 8 of the Agreement on Agriculture by providing export subsidies in excess of its commitment levels as specified in its Schedule. |

*(Continued)*

| Resolution year | Case | Dispute date and status | Parties | | Agreements cited | Summary |
|---|---|---|---|---|---|---|
| | | | Complainant | Respondent | | |
| | | | | | | • The Panel erred in not ruling on the Complaining Parties' claims under the SCM Agreement, because the Panel's ruling under the Agreement on Agriculture was insufficient to fully resolve the dispute, especially in relation to implementation of a remedy; but that because there was insufficient material before it, it was not in a position to complete the legal analysis and to examine the Complaining Parties' claims under the SCM Agreement that were left unaddressed by the Panel. |
| | | | | | | • At its meeting of 19 May 2005, the DSB adopted the Appellate Body report and the Panel report, as modified by the Appellate Body report. |

(*Continued*)

| Resolution year | Case | Parties | | | Agreements cited | Summary |
|---|---|---|---|---|---|---|
| | | Dispute date and status | Complainant | Respondent | | |
| 2005 | DISPUTE DS266 European Communities—Export Subsidies on Sugar | Report(s) adopted, with recommendation to bring measure(s) into conformity on 19 May 2005 (up-to-date at 24 February 2010) | Brazil | European Communities **Third parties:** Australia; Barbados; Belize; Canada; China; Colombia; Cuba; Fiji; Guyana; India; Jamaica; Kenya; Madagascar; Malawi; Mauritius; New Zealand; Paraguay; Saint Kitts and Nevis; Swaziland; Tanzania; Thailand; Trinidad and Tobago; United States; Côte d'Ivoire | Agriculture: Art. 3.3, 8, 9.1, 10.1 GATT 1994: Art. III:4, XVI Subsidies and Countervailing Measures: Art. 1.1, 3.1, 3.2 | **Measure at issue:** EC measures relating to subsidization of the sugar industry, namely, a Common Organization for Sugar (CMO) (set out in Council Regulation (EC) No. 1260/2001): two categories of production quotas—"A sugar" and "B sugar"—were established under the Regulation. Further, sugar produced in excess of A and B quota levels is called C sugar, which is not eligible for domestic price support or direct export subsidies and must be exported. EC export subsidy commitment levels for sugar: The Appellate Body upheld the Panel's finding that footnote 1 in the EC Schedule relating to preferential imports from certain ACP countries and India did not have the legal effect of enlarging or otherwise modifying the European Communities' quantity commitment level contained in Section II, Part IV of its Schedule. |

(*Continued*)

| Resolution year | Case | Dispute date and status | Parties | | Agreements cited | Summary |
| | | | Complainant | Respondent | | |
| --- | --- | --- | --- | --- | --- | --- |
| | | | | | | AA Arts. 9.1(c), 3.3 and 8 (export subsidies—exports of C sugar): The Appellate Body upheld the Panel's finding that the European Communities violated Arts. 3.3 and 8 by exporting C sugar because export subsidies in the form of payments on the export financed by virtue of government action within the meaning of Art. 9.1(c) were provided in excess of the European Communities' commitment level. In this regard, the European Communities provided two types of "payments" within the meaning of Art. 9.1(c) for C sugar producers, i.e., (1) sales of C beet below the total costs of production to C sugar producers; and (2) transfers of financial resources, through cross-subsidization resulting from the operation of the EC sugar regime. Further, the Panel concluded that the European Communities had not demonstrated, pursuant to AA Art. 10.3, that exports of C sugar that exceeded the European Communities' commitment levels since 1995 had not been subsidized. |
| | | | | | | *(Continued)* |

| Resolution year | Case | Dispute date and status | Parties Complainant | Parties Respondent | Agreements cited | Summary |
|---|---|---|---|---|---|---|
| | | | | | | AA Arts. 9.1(a), 3 and 8 (export subsidies—export of ACP/India equivalent sugar): The Panel found that the European Communities acted inconsistently with Arts. 3 and 8 since the evidence indicated that European Communities' exports of ACP/India equivalent sugar received export subsidies within the meaning of Art. 9.1(a) and the European Communities had not proved otherwise. See http://www.wto.org/english/tratop_e/dispu_e/cases_e/1pagesum_e/ds266sum_e.pdf |
| 2009 | DISPUTE DS267 United States—Subsidies on Upland Cotton | Authorization to retaliate granted on 19 November 2009 (up-to-date at 20 November 2012) | Brazil | United States **Third parties:** Argentina; Australia; Benin; Canada; Chad; China; Chinese Taipei; European Communities; India; New Zealand; Pakistan; Paraguay; Venezuela, Bolivarian Republic of; Japan; Thailand | Agriculture: Art. 3.3, 7.1, 8, 9.1, 10.1 GATT 1994: Art. III:4, XVI Subsidies and Countervailing Measures: Art. 3, 5, 6 | **Measure at issue:** US agricultural "domestic support" measures, export credit guarantees and other measures alleged to be export and domestic content subsidies. US export credit guarantees and agricultural domestic support measures relating to cotton, pig meat, poultry meat, and other agricultural products. **Product at issue:** Upland cotton and other products covered by export credit guarantees. See http://www.wto.org/english/tratop_e/dispu_e/cases_e/1pagesum_e/ds267sum_e.pdf |

*(Continued)*

| Resolution year | Case | Dispute date and status | Parties Complainant | Parties Respondent | Agreements cited | Summary |
|---|---|---|---|---|---|---|
| 2003 | DISPUTE DS270 Australia—Certain Measures Affecting the Importation of Fresh Fruit and Vegetables | Panel established, but not yet composed on 29 August 2003 (up-to-date at 24 February 2010) | Philippines | Australia **Third parties:** Chile; China; European Communities; Ecuador; India; Thailand; United States | GATT 1994: Art. XI, XI:1, XIII Import Licensing: Art. 1, 3, 3.2, 3.5 Sanitary and Phytosanitary Measures (SPS): Art. 2, 2.2, 2.3, 3, 3.1, 4, 5, 5.1, 5.2, 5.3, 5.5, 5.6, 6, 6.1, 6.2, 10 | On 18 October 2002, the Philippines requested consultations with Australia on certain measures affecting the importation into Australia of fresh fruit and vegetables, including bananas, which include, but are not limited to: • Section 64 of Quarantine Proclamation 1998 promulgated under the Quarantine Act 1908. • Regulations, requirements, and procedures issued pursuant thereto. • Amendments to any of the foregoing. • Their application. • The Philippines considered that these measures are inconsistent with the obligations of Australia under the GATT 1994, the SPS Agreement and the Agreement on Import Licensing Procedures.... |

*(Continued)*

| Resolution year | Case | Dispute date and status | Parties Complainant | Parties Respondent | Agreements cited | Summary |
|---|---|---|---|---|---|---|
| 2002 | DISPUTE DS271 Australia—Certain Measures Affecting the Importation of Fresh Pineapple | In consultations on 18 October 2002 (up-to-date at 24 February 2010) | | | | On 18 October, [2002], the Philippines requested consultations with Australia on certain measures affecting the importation into Australia of fresh pineapple, which include, but are not limited to: • Section 64 of the Quarantine Proclamation 1998 promulgated under the Quarantine Act 1908. • Regulations, requirements, and procedures issued pursuant thereto, including Plant Biosecurity Policy Memorandum 2002/45 (requiring that fresh pineapple fruit from the Philippines, the Solomon Islands, Sri Lanka, and Thailand shall, among other requirements, be de-crowned and subjected to pre-shipment methyl bromide fumigation as conditions for importation into Australia). • Amendments to any of the foregoing. • Their application. • The Philippines considered that these measures are inconsistent with the obligations of Australia under the GATT 1994 and the SPS Agreement. |

(*Continued*)

| Resolution year | Case | Dispute date and status | Parties | | Agreements cited | Summary |
| | | | Complainant | Respondent | | |
| --- | --- | --- | --- | --- | --- | --- |
| 2002 | DISPUTE DS275 Venezuela—Import Licensing Measures on Certain Agricultural Products | In consultations on 7 November 2002 (up-to-date at 24 February 2010) | United States | Venezuela, Bolivarian Republic of | Agriculture: Art. 4.2 GATT 1994: Art. III, X, XI, XIII Import Licensing: Art. 1.4, 3.2, 3.5, 5.1, 5.2, 5.3 Trade-Related Investment Measures (TRIMs): Art. 2.1 | On 7 November 2002, the United States requested consultations with Venezuela concerning Venezuelan import licensing systems and practices that restrict agricultural imports from the United States. According to the United States, Venezuela has established import licensing requirements for numerous agricultural products, including corn, sorghum, dairy products (e.g., cheese, whey, whole milk powder, and non-fat dry milk), grapes, yellow grease, poultry, beef, pork, and soybean meal. The United States claimed that Venezuela appears to have established a discretionary import licensing regime for the above products. It further claimed that, through its import licensing practices, Venezuela has failed to establish a transparent and predictable system for issuing import licenses and has severely restricted and distorted trade in these goods. (*Continued*) |

| Resolution year | Case | Dispute date and status | Parties | | Agreements cited | Summary |
|---|---|---|---|---|---|---|
| | | | Complainant | Respondent | | |
| | | | | | | The United States alleges that Venezuela's import licensing systems and practices appear to be inconsistent with Article 4.2 of the Agreement on Agriculture; Articles III, X, XI, and XIII of GATT 1994; Article 2.1 of the TRIMs Agreement, and Articles 1.4, 3.2, 3.5, 5.1, 5.2, and 5.3 of the Import Licensing Agreement. Venezuela's measures also appear to nullify or impair the benefits accruing to the United States directly or indirectly under the cited agreements. On 20 November 2002, the EC and Canada requested to join the consultations. On 21 November 2002, New Zealand and Chile requested to join the consultations. On 22 November 2002, Argentina and Colombia requested to join the consultations. On 25 November 2002, Venezuela informed the DSB that it had accepted the requests of Argentina, Canada, Chile, the EC, and New Zealand to join the consultations. |

*(Continued)*

| Resolution year | Case | Parties | | Dispute date and status | Agreements cited | Summary |
|---|---|---|---|---|---|---|
| | | Complainant | Respondent | | | |
| 2002 | DISPUTE DS276 Canada—Measures Relating to Exports of Wheat and Treatment of Imported Grain | United States | Canada **Third parties:** Australia; Chile; China; Chinese Taipei; European Communities; Japan; Mexico | Implementation notified by respondent on 31 August 2005 (up-to-date at 24 February 2010) | GATT 1994: Art. III, III:4, XVII, XVII:1 Trade-Related Investment Measures (TRIMs): Art. 2 | On 17 December 2002, the United States requested consultations with Canada in regards to matters concerning the export of wheat by the Canadian Wheat Board and the treatment accorded by Canada to grain imported into Canada. According to the United States, the actions of the Government of Canada and the Canadian Wheat Board (entity enjoying exclusive rights to purchase and sell Western Canadian wheat for human consumption) related to export of wheat appear to be inconsistent with paragraphs 1(a) and 1(b) of Article XVII of GATT 1994. As regards the treatment of grain imported into Canada, the United States maintains that the following Canadian measures are inconsistent with Article III of the GATT 1994 and Article 2 of TRIMs since they discriminate against imported grain: |

*(Continued)*

| Resolution year | Case | Dispute date and status | Parties | | Agreements cited | Summary |
| --- | --- | --- | --- | --- | --- | --- |
| | | | Complainant | Respondent | | |
| | | | | | | Under the Canadian Grain Act and Canadian regulations, imported wheat cannot be mixed with Canadian domestic grain being received into or discharged out of grain elevators, and Canadian Law caps the maximum revenues that railroads may receive on the shipment of domestic grain, but not revenues received on the shipment of imported grain; Canada provides a preference for domestic grain over imported grain when allocating government-owned railcars. |

*(Continued)*

| Resolution year | Case | Parties — Complainant | Parties — Respondent | Dispute date and status | Agreements cited | Summary |
|---|---|---|---|---|---|---|
| 2005 | DISPUTE DS283 European Communities—Export Subsidies on Sugar | Thailand | EU **Third parties:** Australia; Barbados; Belize; Brazil; Canada; China; Colombia; Cuba; Fiji; Guyana; India; Jamaica; Kenya; Madagascar; Malawi; Mauritius; New Zealand; Paraguay; Saint Kitts and Nevis; Swaziland; Tanzania; Trinidad and Tobago; United States; Côte d'Ivoire | Report(s) adopted, with recommendation to bring measure(s) into conformity on 19 May 2005 (up-to-date at 24 February 2010) | Agriculture: Art. 3.3, 8, 9.1, 10.1 GATT 1994: Art. III:4 Subsidies and Countervailing Measures: Art. 3, 3.1, 3.2 | **Measure at issue:** EC measures relating to subsidization of the sugar industry, namely, a Common Organization for Sugar (CMO) (set out in Council Regulation (EC) No. 1260/2001): two categories of production quotas—"A sugar" and "B sugar"—were established under the Regulation. Further, sugar produced in excess of A and B quota levels is called C sugar, which is not eligible for domestic price support or direct export subsidies and must be exported. **Industry at Issue:** Sugar industry See http://www.wto.org/english/tratop_e/dispu_e/cases_e/1pagesum_e/ds283sum_e.pdf |

*(Continued)*

| Resolution year | Case | Dispute date and status | Parties | | Agreements cited | Summary |
|---|---|---|---|---|---|---|
| | | | Complainant | Respondent | | |
| 2004 | DISPUTE DS284 Mexico—Certain Measures Preventing the Importation of Black Beans from Nicaragua | Settled or terminated (withdrawn, mutually agreed solution) on 8 March 2004 (up-to-date at 24 February 2010) | Nicaragua | Mexico | GATT 1994: Art. I, X, XI, XIII Import Licensing: Art. 1, 2.2(a) Sanitary and Phytosanitary Measures (SPS): Art. 2, 5.1, 7, Annex B | On 17 March 2003, Nicaragua requested consultations with Mexico regarding certain measures imposed by Mexico, which prevent the importation of black beans from Nicaragua. These measures include: The administration of the procedures set out in Official Standard 006-FITO-95 and Official Standard 028-FITO-95, including the refusal of the competent Mexican authorities to furnish importers with the document containing the phytosanitary requirements necessary for the importation of black beans from Nicaragua. The more favorable treatment that the competent Mexican authorities accord in the administration of the above procedures to like products originating in countries other than Nicaragua. The failure to publish the specific phytosanitary requirements for the importation of black beans from Nicaragua. |

*(Continued)*

| Resolution year | Case | Dispute date and status | Parties | | Agreements cited | Summary |
|---|---|---|---|---|---|---|
| | | | Complainant | Respondent | | |
| | | | | | | The failure to publish the rules, requirements and procedures concerning the tender for the quota allocation of black beans from Nicaragua. On 8 March 2004, Nicaragua informed the DSB that it wished formally to withdraw the request for consultations as its complaints had been adequately addressed as a result of negotiations with Mexico. |

*(Continued)*

| Resolution year | Case | Dispute date and status | Parties | | Agreements cited | Summary |
|---|---|---|---|---|---|---|
| | | | Complainant | Respondent | | |
| 2007 | DISPUTE DS287 Australia—Quarantine Regime for Imports | Settled or terminated (withdrawn, mutually agreed solution) on 9 March 2007 (up-to-date at 24 February 2010) | EU | Australia **Third parties:** Canada; Chile; China; India; Philippines; Thailand; United States | Sanitary and Phytosanitary Measures (SPS): Art. 2.2, 2.3, 3.3, 4.1, 5.1, 5.6, 5.7, 8, Annex C | On 3 April 2003, the EC requested consultations with Australia regarding the Australian quarantine regime for imports, both as such and as applied to certain specific cases. According to the EC, the Australian quarantine regime for imports appears to be governed both by legislation as well as by the exercise of discretion granted to a Director of Quarantine and by administrative guidance issued on the exercise of that discretion. As regards the quarantine regime as such, the EC claims that the effect of this regime appears to be that the import of products is *a priori* prohibited, although there is no risk assessment. Risk assessments appear to be commenced, if at all, only once the import of a product has been specifically requested. In some cases, no risk assessment has been commenced despite such request. In other cases it has been commenced but not completed. As regards specific cases, the EC claims that: |

*(Continued)*

| Resolution year | Case | Dispute date and status | Parties | | Agreements cited | Summary |
| --- | --- | --- | --- | --- | --- | --- |
| | | | Complainant | Respondent | | |
| | | | | | | Australia permits the import of deboned pig meat from Denmark for processing in Australia, but refuses the import of processed deboned pig meat from Denmark. It also claims that the processing requirements imposed in Australia may be more trade-restrictive than necessary in the circumstances to protect Australia from PRRS (Porcine Reproductive and Respiratory Syndrome). It also appears that requests have been made for access to Australia for processed pig meat or deboned pig meat for processing from other EU Member States, which have been refused. |

*(Continued)*

| Resolution year | Case | Dispute date and status | Parties Complainant | Parties Respondent | Agreements cited | Summary |
|---|---|---|---|---|---|---|
| | | | | | | Australia permits the import of poultry meat which has been cooked to high temperature and for long periods to prevent the entry of IBD (infectious bursal disease). The EC claims that it appears that IBD may already be present in the Australian poultry flock and that no efforts are being made to eradicate it. The EC also claims that the processing requirements imposed in Australia may be more trade-restrictive than necessary in the circumstances to protect Australia from IBD. |

*(Continued)*

| Resolution year | Case | Dispute date and status | Parties | | Agreements cited | Summary |
|---|---|---|---|---|---|---|
| | | | Complainant | Respondent | | |
| 2008 | DISPUTE DS291 European Communities—Measures Affecting the Approval and Marketing of Biotech Products (See also DS292 and DS293) | Authorization to retaliate requested (including 22.6 arbitration) on 17 January 2008 (up-to-date at 24 February 2010) | United States | EU **Third parties:** Argentina; Australia; Brazil; Canada; Chile; China; Chinese Taipei; Colombia; El Salvador; Honduras; Mexico; New Zealand; Norway; Paraguay; Peru; Thailand; Uruguay | Agriculture: Art. 4, 4.2 GATT 1994: Art. I, I:1, III, III:4, X, X:1, XI, XI:1 Sanitary and Phytosanitary Measures (SPS): Art. 2, 2.2, 2.3, 5, 5.1, 5.2, 5.5, 5.6, 7, 8, Annex B, Annex C Technical Barriers to Trade (TBT): Art. 2, 2.1, 2.2, 2.8, 2.9, 2.11, 2.12, 5, 5.1, 5.2, 5.6, 5.8 | **Measure at issue:** (1) Alleged general EC moratorium on approvals of biotech products; (2) EC measures allegedly affecting the approval of specific biotech products; and (3) EC member State safeguard measures prohibiting the import/marketing of specific biotech products within the territories of these member States. **Product at issue:** Agricultural biotech products from the United States, Canada, and Argentina |

*(Continued)*

| Resolution year | Case | Dispute date and status | Parties | | Agreements cited | Summary |
|---|---|---|---|---|---|---|
| | | | Complainant | Respondent | | |
| | | | | | | On 13 May 2003, Canada requested consultations with the EC concerning certain measures taken by the EC and its member States affecting imports of agricultural and food imports from Canada. Regarding EC-level measures, Canada asserted that the moratorium applied by the EC since October 1998 on the approval of biotech products has restricted imports of agricultural and food products from Canada. Regarding member State-level measures, Canada asserted that a number of EC member States maintain national marketing and import bans on biotech products even though those products have already been approved by the EC for import and marketing in the EC. |

*(Continued)*

| Resolution year | Case | Dispute date and status | Parties | | Agreements cited | Summary |
| | | | Complainant | Respondent | | |
| --- | --- | --- | --- | --- | --- | --- |
| 2009 | DISPUTE DS292 European Communities—Measures Affecting the Approval and Marketing of Biotech Products | Settled or terminated (withdrawn, mutually agreed solution) on 15 July 2009 (up-to-date at 24 February 2010) | Canada | EU **Third parties:** Argentina; Australia; Brazil; Chile; China; Chinese Taipei; Colombia; El Salvador; Honduras; Mexico; New Zealand; Norway; Paraguay; Peru; Thailand; Uruguay; United States | Agriculture: Art. 4.2 GATT 1994: Art. I:1, III:4, X:1, XI:1 Sanitary and Phytosanitary Measures (SPS): Art. 2, 2.2, 2.3, 5, 5.1, 5.2, 5.5, 5.6, 7, 8, Annex B, Annex C Technical Barriers to Trade (TBT): Art. 2, 2.1, 2.2, 2.8, 2.9, 2.11, 2.12, 5, 5.1, 5.2, 5.6, 5.8 | |

*(Continued)*

| Resolution year | Case | Dispute date and status | Parties | | Agreements cited | Summary |
|---|---|---|---|---|---|---|
| | | | Complainant | Respondent | | |
| 2010 | DISPUTE DS293 European Communities—Measures Affecting the Approval and Marketing of Biotech Products | Settled or terminated (withdrawn, mutually agreed solution) on 19 March 2010 | Argentina | EU **Third parties:** Australia; Brazil; Canada; Chile; China; Chinese Taipei; Colombia; El Salvador; Honduras; Mexico; New Zealand; Norway; Paraguay; Peru; Thailand; Uruguay; United States | Agriculture: Art. 4, 4.2 GATT 1994: Art. I, I:1, III, III:4, X, X:1, XI, XI:1, X:3(a) Sanitary and Phytosanitary Measures (SPS): Art. 2, 2.2, 2.3, 5, 5.1, 5.2, 5.5, 5.6, 7, 8, 10, 10.1, Annex B, Annex C Technical Barriers to Trade (TBT): Art. 2, 2.1, 2.2, 2.8, 2.9, 2.11, 5, 5.1, 5.2, 5.6, 5.8, 12 | |

*(Continued)*

| Resolution year | Case | Dispute date and status | Parties | | Agreements cited | Summary |
| --- | --- | --- | --- | --- | --- | --- |
| | | | Complainant | Respondent | | |
| 2005 | DISPUTE DS295 Mexico—Definitive Anti-Dumping Measures on Beef and Rice | Report(s) adopted, with recommendation to bring measure(s) into conformity on 20 December 2005 (up-to-date at 24 February 2010) | United States | Mexico **Third parties:** China; European Communities; Turkey | Anti-dumping (Article VI of GATT 1994): Art. 1, 3, 3.1, 3.2, 3.4, 3.5, 4.1, 5.8, 6, 6.1, 6.1.1, 6.2, 6.4, 6.8, 6.9, 6.10.7, 9, 9.3, 9.4, 9.5, 10.6, 11, 11.1, 11.9, 12, 12.1, 12.2, 18.1, 19.3, 21.1, 21.2, 32.1, Annex II GATT 1994: Art. VI, VI:2 Subsidies and Countervailing Measures: Art. 11.9, 12.1.1, 12.5, 12.7, 17, 19, 19.3, 20.6, 21, 21.1, 21.2, 32.1 | On 16 June 2003, the United States requested consultations with Mexico concerning its definitive anti-dumping measures on beef and long grain white rice, as well as certain provisions of Mexico's Foreign Trade Act and its Federal Code of Civil Procedure. The United States claimed that these measures were inconsistent with Mexico's obligations under the provisions of GATT 1994, the Anti-Dumping Agreement and the SCM Agreement: Long-grain white rice from the United States. |

*(Continued)*

| Resolution year | Case | Dispute date and status | Parties | | Agreements cited | Summary |
|---|---|---|---|---|---|---|
| | | | Complainant | Respondent | | |
| 2006 | DISPUTE DS323 Japan—Import Quotas on Dried Laver and Seasoned Laver | Settled or terminated (withdrawn, mutually agreed solution) on 23 January 2006 (up-to-date at 24 February 2010) | Korea, Republic of | Japan **Third Parties:** China; European Communities; New Zealand; United States | Agriculture: Art. 4.2 GATT 1994: Art. X:3, XI, X:3(a) Import Licensing: Art. 1.2, 1.6 | On 1 December 2004, Korea requested consultations with Japan concerning Japan's import quotas on dried laver and seasoned laver. According to the request for consultations, Korea believes that Japan's extremely restrictive import quotas on dried laver and seasoned laver are inconsistent with, *inter alia*: Articles X.3 and XI of the GATT 1994. Article 4.2 of the Agreement on Agriculture. Article 1.2 and 1.6 of the Agreement on Import Licensing Procedures. *(Continued)* |

| Resolution year | Case | Dispute date and status | Parties | | Agreements cited | Summary |
|---|---|---|---|---|---|---|
| | | | Complainant | Respondent | | |
| 2006 | DISPUTE DS338 Canada—Provisional Anti-Dumping and Countervailing Duties on Grain Corn from the United States | In consultations on 17 March 2006 (up-to-date at 24 February 2010) | United States | Canada | Anti-dumping (Article VI of GATT 1994): Art. 1, 3, 3.2, 3.4, 3.5, 7, 12.2.1 GATT 1994: Art. VI Subsidies and Countervailing Measures: Art. 10, 15, 15.2, 15.4, 15.5, 17, 22.4 | On 17 March 2006, the United States requested consultations with Canada concerning Canada's imposition of provisional anti-dumping and countervailing duties on unprocessed grain corn from the United States published in the Canada Gazette on 31 December 2005, and also with respect to certain provisions of Canada's Special Import Measures Act. The United States believes that these provisional duties are inconsistent with Canada's obligations under provisions of the GATT 1994, the Anti-Dumping Agreement and the SCM Agreement. These include Article 3 of the Anti-Dumping Agreement and Article 15 of the SCM Agreement with respect to a number of factors relating to the preliminary injury determination in this case and Articles 1, 7, and 12.2.1 of the Anti-Dumping Agreement, Articles 10, 17, and 22.4 of the SCM Agreement, and Article VI of GATT 1994 by virtue of Canada having imposed the said provisional duties based, in part, on a WTO-inconsistent preliminary injury determination. |

*(Continued)*

| Resolution year | Case | Dispute date and status | Parties | | Agreements cited | Summary |
|---|---|---|---|---|---|---|
| | | | Complainant | Respondent | | |
| 2007 | DISPUTE DS365 United States—Domestic Support and Export Credit Guarantees for Agricultural Products | Panel established, but not yet composed on 17 December 2007 | Brazil | United States **Third parties:** Argentina; Australia; Chile; European Union; India; Japan; Mexico; New Zealand; Nicaragua; South Africa; Chinese Taipei; Thailand; Turkey; Uruguay | Agriculture: Art. 3.2, 3.3, 6.1, 8, 9.1, 10.1 Subsidies and Countervailing Measures: Art. 3.1, 3.2 | On 8 January 2007, Canada requested consultations with the United States concerning three different categories of measures. First, Canada claims that the United States provides subsidies to the US corn industry that are specific to US producers of primary agricultural products and/or to the US corn industry. Canada considers that the measures at issue are inconsistent with Articles 5(c) and 6.3(c) of the SCM Agreement. Second, Canada claims that the United States makes available to its exporters premium rates and other terms more favorable than those which the market would otherwise provide through export credit guarantee programs under the Agricultural Trade Act of 1978 and other measures, such as the GSM-102 program and SCGP, as well as the programs, legislation, regulations, and statutory instruments providing the support. |

*(Continued)*

| Resolution year | Case | Dispute date and status | Parties | | Agreements cited | Summary |
|---|---|---|---|---|---|---|
| | | | Complainant | Respondent | | Canada considers that these programs provide subsidies contingent upon export performance contrary to Article 3.1(a) and 3.2 of the SCM Agreement, and they also violate Articles 3.3, 8, 9.1 and 10.1 of the Agreement on Agriculture. Third, Canada claims that, through the improper exclusion of domestic support, the United States provides support in favor of domestic producers in excess of the commitment levels specified in Section I of Part IV of the Schedule, contrary to Article 3.2 of the Agreement on Agriculture. *(Continued)* |

| Resolution year | Case | Dispute date and status | Parties | | Agreements cited | Summary |
| --- | --- | --- | --- | --- | --- | --- |
| | | | Complainant | Respondent | | |
| 2007 | DISPUTE DS365 United States—Domestic Support and Export Credit Guarantees for Agricultural Products | Panel established, but not yet composed on 17 December 2007 (up-to-date at 24 February 2010) | Brazil | United States **Third parties:** Argentina; Australia; Chile; European Union; India; Japan; Mexico; New Zealand; Nicaragua; South Africa; Chinese Taipei; Thailand; Turkey; Uruguay | Agriculture: Art. 3.2, 3.3, 6.1, 8, 9.1, 10.1 Subsidies and Countervailing Measures: Art. 3.1, 3.2 | On 8 January 2007, Canada requested consultations with the United States concerning three different categories of measures. First, Canada claims that the United States provides subsidies to the US corn industry that are specific to US producers of primary agricultural products and/or to the US corn industry. Canada considers that the measures at issue are inconsistent with Articles 5(c) and 6.3(c) of the SCM Agreement. Second, Canada claims that the United States makes available to its exporters premium rates and other terms more favorable than those which the market would otherwise provide through export credit guarantee programs under the Agricultural Trade Act of 1978 and other measures such as the GSM-102 program and SCGP as well as the programs, legislation, regulations and statutory instruments providing the support. Canada considers that these programs provide subsidies contingent upon export performance contrary to Article 3.1(a) and 3.2 of the SCM Agreement, and they also violate Articles 3.3, 8, 9.1 and 10.1 of the Agreement on Agriculture. *(Continued)* |

| Resolution year | Case | Dispute date and status | Parties Complainant | Parties Respondent | Agreements cited | Summary |
|---|---|---|---|---|---|---|
| | | | | | | Third, Canada claims that, through the improper exclusion of domestic support, the United States provides support in favor of domestic producers in excess of the commitment levels specified in Section I of Part IV of the Schedule, contrary to Article 3.2 of the Agreement on Agriculture. |
| 2013 | DISPUTE DS430 India—Measures Concerning the Importation of Certain Agricultural Products from the United States | Panel composed on 18 February 2013 (up-to-date at 13 March 2013) | United States | India **Third parties:** China; Colombia; Ecuador; European Union; Guatemala; Japan; Viet Nam; Argentina; Australia; Brazil | Sanitary and Phytosanitary Measures (SPS): Art. 2, 2.2, 2.3, 3.1, 5, 5.1, 5.2, 5.5, 5.6, 5.7, 6, 6.1, 6.2, 7, Annex B GATT 1994: Art. I, XI | On 6 March 2012, the United States requested consultations with India with respect to the prohibitions imposed by India on the importation of various agricultural products from the United States purportedly because of concerns related to Avian Influenza. The measures at issue are: the Indian Livestock Importation Act, 1898 (9 of 1898) ("Livestock Act"); a number of orders issued by India's Department of Animal Husbandry, Dairying, and Fisheries pursuant to the Livestock Act, most recently S.O. 1663(E); as well as any amendments, related measures, or implementing measures. |

*(Continued)*

| Resolution year | Case | Dispute date and status | Parties | | Agreements cited | Summary |
|---|---|---|---|---|---|---|
| | | | Complainant | Respondent | | |
| 2013 | DISPUTE DS457 Peru—Additional Duty on Imports of Certain Agricultural Products | In consultations on 12 April 2013 (up-to-date at 12 April 2013) | Guatemala | Peru | Agriculture: Art. 4.2 GATT 1994: Art. II:1, II:1(b), X:1, X:3(a), XI, XI:1 Customs valuation (Article VII of GATT 1994): Art. 1, 2, 3, 5, 6, 7 | On 12 April 2013, Guatemala requested consultations with Peru with respect to the imposition by Peru of an "additional duty" on imports of certain agricultural products, such as rice, sugar, maize, milk, and certain dairy products. |
| 1995 | DISPUTE DS5 Korea—Measures Concerning the Shelf-Life of Products | Settled or terminated (withdrawn, mutually agreed solution) on 20 July 1995 (up-to-date at 24 February 2010) | United States | Korea, Republic of | Agriculture: Art. 4 GATT 1994: Art. III, XI Sanitary and Phytosanitary Measures (SPS): Art. 2, 5 Technical Barriers to Trade (TBT): Art. 2 | On 3 May 1995, the United States requested consultations with Korea in respect of requirements imposed by Korea on imports from the United States, which had the effect of restricting imports. The United States alleged violations of Articles III and XI of GATT, Articles 2 and 5 of the SPS Agreement, Article 2 of the TBT Agreement, and Article 4 of the Agreement on Agriculture. |

*(Continued)*

| Resolution year | Case | Dispute date and status | Parties | | Agreements cited | Summary |
|---|---|---|---|---|---|---|
| | | | Complainant | Respondent | | |
| 1996 | DISPUTE DS7 European Communities—Trade Description of Scallops | Settled or terminated (withdrawn, mutually agreed solution) on 5 July 1996 (up-to-date at 24 February 2010) | Canada | EU **Third parties:** Australia; Chile; Iceland; Japan; Peru; United States | GATT 1994: Art. I, III Technical Barriers to Trade (TBT): Art. 2 | Complaints by Canada, Peru and Chile. The complaint concerned a French Government Order laying down the official name and trade description of scallops. Complainants claimed that this Order will reduce competitiveness on the French market as their product will no longer be able to be sold as "Coquille Saint-Jacques" although there is no difference between their scallops and French scallops in terms of color, size, texture, appearance, and use, i.e., it is claimed they are "like products." Violations of GATT Articles I and III and TBT Article 2 were alleged. |

*(Continued)*

| Resolution year | Case | Dispute date and status | Parties | | Agreements cited | Summary |
| --- | --- | --- | --- | --- | --- | --- |
| | | | Complainant | Respondent | | |
| 1995 | DISPUTE DS9 European Communities—Duties on Imports of Cereals | Panel established, but not yet composed on 11 October 1995 (up-to-date at 24 February 2010) | Canada | EU | GATT 1994: Art. II, VII Customs valuation (Article VII of GATT 1994): Art. 1 | Canada requested consultations with the EC on 10 July 1995 concerning EC regulations implementing some of the EC's Uruguay Round concessions on agriculture, specifically, regulations which impose a duty on wheat imports based on reference prices rather than transaction values, with the result that the duty-paid import price for Canadian wheat will be greater than the effective intervention price increased by 55% whenever the transaction value is greater than the representative price. *(Continued)* |

| Resolution year | Case | Dispute date and status | Parties Complainant | Parties Respondent | Agreements cited | Summary |
|---|---|---|---|---|---|---|
| 1996 | DISPUTE DS12 European Communities—Trade Description of Scallops | Settled or terminated (withdrawn, mutually agreed solution) on 5 July 1996 (up-to-date at 24 February 2010) | Peru | EU **Third parties:** Australia; Canada; Iceland; Japan; United States | GATT 1994: Art. I, III Technical Barriers to Trade (TBT): Art. 2, 2.1, 2.2, 12 | The complaint concerned a French Government Order laying down the official name and trade description of scallops. Complainants claimed that this Order will reduce competitiveness on the French market as their product will no longer be able to be sold as "Coquille Saint-Jacques," although there is no difference between their scallops and French scallops in terms of color, size, texture, appearance, and use, i.e., it is claimed they are "like products." Violations of GATT Articles I and III and TBT Article 2 were alleged. |
| 1997 | DISPUTE DS13 European Communities—Duties on Imports of Grains | Settled or terminated (withdrawn, mutually agreed solution) on 2 May 1997 (up-to-date at 24 February 2010) | United States | EU | GATT 1994: Art. I, II, VII, X Customs valuation (Article VII of GATT 1994): Art. 1, 2, 3, 4, 5, 6, 7, 8, 9, 11, 22, Annex I | This request for consultations, dated 19 July 1995, has potentially broader product coverage than the case brought by Canada (WT/DS9) but otherwise concerns much the same issues. |

*(Continued)*

| Resolution year | Case | Dispute date and status | Parties | | Agreements cited | Summary |
| --- | --- | --- | --- | --- | --- | --- |
| | | | Complainant | Respondent | | |
| 1996 | DISPUTE DS14 European Communities—Trade Description of Scallops | Settled or terminated (withdrawn, mutually agreed solution) on 5 July 1996 (up-to-date at 24 February 2010) | Chile | EU | GATT 1994: Art. I, III Technical Barriers to Trade (TBT): Art. 2.1, 2.2 | The complaint concerned a French Government Order laying down the official name and trade description of scallops. Complainants claimed that this Order will reduce competitiveness on the French market as their product will no longer be able to be sold as "Coquille Saint-Jacques," although there is no difference between their scallops and French scallops in terms of color, size, texture, appearance, and use, i.e., it is claimed they are "like products." Violations of GATT Articles I and III and TBT Article 2 were alleged. |
| 2012 | DISPUTE DS16 European Communities—Regime for the Importation, Sale and Distribution of Bananas | Settled or terminated (withdrawn, mutually agreed solution) on 8 November 2012 | Guatemala; Honduras; Mexico; United States | EU | Services (GATS): Art. II, XVI, XVII GATT 1994: Art. I, II, III, X, XIII Import Licensing: Art. 1, 3 | On 28 September 1995, Guatemala, Honduras, Mexico, and the United States requested consultations with the European Communities concerning the EC regime for the importation, sale, and distribution of bananas. The EC measures are alleged to be inconsistent with Articles I, II, III, X, and XIII of GATT 1994, Articles 1 and 3 of the Import Licensing Agreement, and Articles II, XVI, and XVII of GATS. |

*(Continued)*

| Resolution year | Case | Dispute date and status | Parties | | Agreements cited | Summary |
|---|---|---|---|---|---|---|
| | | | Complainant | Respondent | | |
| | DISPUTE DS17 European Communities—Duties on Imports of Rice | In consultations on 3 October 1995 (up-to-date at 24 February 2010) | Thailand | EU | GATT 1994: Art. I, II, VII Customs valuation (Article VII of GATT 1994): Art. 1, 2, 3, 4, 5, 6, 7, Annex I | This request for consultations, dated 3 October 1995, covers more or less the same grounds as Canadian (WT/DS9) and the US (WT/DS13) complaints over the EC duties on grains. In addition, Thailand seems to have alleged that the EC has violated the most-favored-nation requirement under GATT Article I in their preferential treatment of basmati rice from India and Pakistan. See also the Uruguayan complaint (WT/DS25). |
| 2000 | DISPUTE DS18 Australia—Measures Affecting Importation of Salmon | Mutually acceptable solution on implementation notified on 18 May 2000 (up-to-date at 24 February 2010) | Canada | Australia **Third parties:** EU; India; Norway; United States | GATT 1994: Art. XI, XIII Sanitary and Phytosanitary Measures (SPS): Art. 2, 3, 5 | On 5 October 1995, Canada requested consultations with Australia in respect of Australia's prohibition of imports of salmon from Canada based on a quarantine regulation. Canada alleged that the prohibition is inconsistent with Articles XI and XIII of the GATT 1994 and also inconsistent with the SPS Agreement. |

*(Continued)*

| Resolution year | Case | Dispute date and status | Parties Complainant | Parties Respondent | Agreements cited | Summary |
|---|---|---|---|---|---|---|
| | | | | | | Australia had acted inconsistently with Articles 5.1 and 2.2 of the SPS Agreement. The Appellate Body broadened the panel's finding that Australia had acted inconsistently with Articles 5.5 and 2.3 of the SPS Agreement. The Appellate Body reversed the panel's finding that Australia had acted inconsistently with Article 5.6 of the SPS Agreement but was unable to come to a conclusion whether or not Australia's measure was consistent with Article 5.6 due to insufficient factual findings by the panel. |
| 1996 | DISPUTE DS20 Korea—Measures concerning Bottled Water | Settled or terminated (withdrawn, mutually agreed solution) on 24 April 1996 (up-to-date at 24 February 2010) | Canada | Korea, Republic of | GATT 1994: Art. III, XI Sanitary and Phytosanitary Measures (SPS): Art. 2, 5 Technical Barriers to Trade (TBT): Art. 2 | In this request for consultations dated 8 November 1995, Canada claimed that Korean regulations on the shelf-life and physical treatment (disinfection) of bottled water were inconsistent with GATT Articles III and XI, SPS Articles 2 and 5 and TBT Article 2. |

*(Continued)*

| Resolution year | Case | Dispute date and status | Parties | | Agreements cited | Summary |
|---|---|---|---|---|---|---|
| | | | Complainant | Respondent | | |
| 2000 | DISPUTE DS21 Australia—Measures Affecting the Importation of Salmonids | Settled or terminated (withdrawn, mutually agreed solution) on 27 October 2000 (up-to-date at 24 February 2010) | United States | Australia **Third parties:** Canada; European Communities; Hong Kong, China; Iceland; India; Norway | GATT 1994: Art. XI Sanitary and Phytosanitary Measures (SPS): Art. 2, 5, 7, 8 | This request for consultations, dated 17 November 1995, concerns the same regulation alleged to be in violation of the WTO Agreements in WT/ DS18, in respect of which the reports of the panel and Appellate Body have already been adopted and are awaiting implementation. |
| 1997 | DISPUTE DS22 Brazil—Measures Affecting Desiccated Coconut | Report(s) adopted, no further action required on 20 March 1997 (up-to-date at 24 February 2010) | Philippines | Brazil **Third parties:** Canada; European Communities; Indonesia; Malaysia; Sri Lanka; United States | Agriculture: Art. 13 GATT 1994: Art. VI:3, VI:6 | On 27 November 1995, the Philippines requested consultations with Brazil in respect of a countervailing duty imposed by Brazil on the Philippine's exports of desiccated coconut. The Philippines claimed that this duty was inconsistent with WTO and GATT rules. |
| 1995 | DISPUTE DS25 European Communities— Implementation of the Uruguay Round Commitments Concerning Rice | In consultations on 12 December 1995 (up-to-date at 24 February 2010) | Uruguay | EU | GATT 1994: Art. XXII:1 | Complaint by Uruguay. This request for consultations, dated 18 December 1995, seems similar to the claim by Thailand (WT/DS17). |

*(Continued)*

| Resolution year | Case | Dispute date and status | Parties | | Agreements cited | Summary |
| --- | --- | --- | --- | --- | --- | --- |
| | | | Complainant | Respondent | | |
| 2009 | DISPUTE DS26 European Communities—Measures Concerning Meat and Meat Products (Hormones) | Mutually acceptable solution on implementation notified on 25 September 2009 (up-to-date at 3 November 2011) | United States | EU **Third parties:** Australia; Canada; New Zealand; Norway | Agriculture: Art. 4 GATT 1994: Art. III, XI Sanitary and Phytosanitary Measures (SPS): Art. 2, 3, 5 Technical Barriers to Trade (TBT): Art. 2 | On 26 January 1996, the United States requested consultations with the European Communities claiming that measures taken by the EC under the Council Directive Prohibiting the Use in Livestock Farming of Certain Substances Having a Hormonal Action restrict or prohibit imports of meat and meat products from the United States and are apparently inconsistent with Articles III or XI of the GATT 1994, Articles 2, 3 and 5 of the SPS Agreement, Article 2 of the TBT Agreement and Article 4 of the Agreement on Agriculture. |

*(Continued)*

| Resolution year | Case | Dispute date and status | Parties | | Agreements cited | Summary |
| --- | --- | --- | --- | --- | --- | --- |
| | | | Complainant | Respondent | | |
| | | | | | | On 25 September 2009, the European Communities and the United States notified the DSB of a Memorandum of Understanding regarding the importation of beef from animals not treated with certain growth-promoting hormones and increased duties applied by the United States to certain products of the European Communities, agreed by the United States and the European Communities on 13 May 2009, in relation to this dispute. |

*(Continued)*

| Resolution year | Case | Dispute date and status | Parties — Complainant | Parties — Respondent | Agreements cited | Summary |
|---|---|---|---|---|---|---|
| 2012 | DISPUTE DS27 European Communities—Regime for the Importation, Sale and Distribution of Bananas | Settled or terminated (withdrawn, mutually agreed solution) on 8 November 2012 (up-to-date at 20 November 2012) | Ecuador; Guatemala; Honduras; Mexico; United States | EU **Third parties:** Belize; Cameroon; Canada; Colombia; Costa Rica; Dominica; Dominican Republic; Ghana; Grenada; India; Jamaica; Japan; Mauritius; Nicaragua; Panama; Philippines; Saint Lucia; Saint Vincent and the Grenadines; Senegal; Suriname; Venezuela, Bolivarian Republic of; Côte d'Ivoire; Brazil; Madagascar | Agriculture: Art. 19 Services (GATS): Art. II, IV, XVI, XVII GATT 1994: Art. I, II, III, X, XI, XIII Import Licensing: Art. 1, 3 Trade-Related Investment Measures (TRIMs): Art. 2, 5 | Complaints by Ecuador, Guatemala, Honduras, Mexico, and the United States. **Product at issue:** Bananas imported from third countries The complainants in this case other than Ecuador had requested consultations with the European Communities on the same issue on 28 September 1995 (DS16). After Ecuador's accession to the WTO, the current complainants again requested consultations with the European Communities on 5 February 1996. The complainants alleged that the European Communities' regime for importation, sale and distribution of bananas is inconsistent with Articles I, II, III, X, XI, and XIII of the GATT 1994, as well as provisions of the Import Licensing Agreement, the Agreement on Agriculture, the TRIMs Agreement and the GATS. |

*(Continued)*

| Resolution year | Case | Dispute date and status | Parties | | Agreements cited | Summary |
|---|---|---|---|---|---|---|
| | | | Complainant | Respondent | | |
| 1996 | DISPUTE DS30 Brazil—Countervailing Duties on Imports of Desiccated Coconut and Coconut Milk Powder from Sri Lanka | In consultations on 23 February 1996 (up-to-date at 24 February 2010) | Sri Lanka | Brazil | GATT 1994: Art. I, II, VI | On 23 February 1996, Sri Lanka requested consultations with Brazil concerning Brazil's imposition of countervailing duties on Sri Lanka's export of desiccated coconut and coconut milk powder. Sri Lanka alleged that those measures are inconsistent with GATT Articles I, II, and VI and Article 13(a) of the Agriculture Agreement (the so-called peace clause). See WT/DS22. |
| 1996 | DISPUTE DS41 Korea—Measures Concerning Inspection of Agricultural Products | In consultations on 24 May 1996 (up-to-date at 24 February 2010) | United States | Korea, Republic of | Agriculture: Art. 4 GATT 1994: Art. III, XI Sanitary and Phytosanitary Measures (SPS): Art. 2, 5, 8 Technical Barriers to Trade (TBT): Art. 2, 5, 6 | On 24 May 1996, the United States requested consultations with Korea concerning testing, inspection, and other measures required for the importation of agricultural products into Korea. The United States claimed that these measures restrict imports and appear to be inconsistent with the WTO Agreement. Violations of GATT Articles III and XI, SPS Articles 2, 5 and 8, TBT Articles 2, 5 and 6, and Article 4 of the Agreement on Agriculture are alleged. The United States requested consultations with Korea on similar issues on 4 April 1995 (WT/DS3). |

*(Continued)*

| Resolution year | Case | Dispute date and status | Parties | | Agreements cited | Summary |
|---|---|---|---|---|---|---|
| | | | Complainant | Respondent | | |
| 2011 | DISPUTE DS48 European Communities—Measures Concerning Meat and Meat Products (Hormones) | Mutually acceptable solution on implementation notified on 17 March 2011 (up-to-date at 3 November 2011) | Canada | EU **Third parties:** Australia; New Zealand; Norway; United States | Agriculture: Art. 4 GATT 1994: Art. III, XI Sanitary and Phytosanitary Measures (SPS): Art. 2, 3, 5 Technical Barriers to Trade (TBT): Art. 2, 5 | Findings: SPS Art. 3.1 (international standards): The Appellate Body rejected the Panel's interpretation and said that the requirement that SPS measures be "based on" international standards, guidelines, or recommendations under Art. 3.1 does not mean that SPS measures must "conform to" such standards. Relationship between SPS Arts. 3.1/3.2 and 3.3 (harmonization): The Appellate Body rejected the Panel's interpretation that Art. 3.3 is the exception to Arts. 3.1 and 3.2 assimilated together and found that Arts. 3.1, 3.2, and 3.3 apply together, each addressing a separate situation. Accordingly, it reversed the Panel's finding that the burden of proof for the violation under Art. 3.3, as a provision providing the exception, shifts to the responding party. |

*(Continued)*

| Resolution year | Case | Dispute date and status | Parties | | Agreements cited | Summary |
| --- | --- | --- | --- | --- | --- | --- |
| | | | Complainant | Respondent | | |
| | | | | | | SPS Art. 5.1 (risk assessment): While upholding the Panel's ultimate conclusion that the EC measure violated Art. 5.1 (and thus Art. 3.3) because it was not based on a risk assessment, the Appellate Body reversed the Panel's interpretation, considering that Art. 5.1 requires that there be a "rational relationship" between the measure at issue and the risk assessment. SPS Art. 5.5 (prohibition on discrimination and disguised restriction on international trade): The Appellate Body reversed the Panel's finding that the EC measure, through arbitrary or unjustifiable distinctions, resulted in "discrimination or a disguised restriction of international trade" in violation of Art. 5.5, noting: (1) the evidence showed that there were genuine anxieties concerning the safety of the hormones; (2) the necessity for harmonizing measures was part of the effort to establish a common internal market for beef; and (3) the Panel's finding was not supported by the "architecture and structure" of the measures. |

*(Continued)*

| Resolution year | Case | Dispute date and status | Parties | | Agreements cited | Summary |
| --- | --- | --- | --- | --- | --- | --- |
| | | | Complainant | Respondent | | |
| 1996 | DISPUTE DS49 United States—Anti-Dumping Investigation Regarding Imports of Fresh or Chilled Tomatoes from Mexico | In consultations on 1 July 1996 (up-to-date at 24 February 2010) | Mexico | United States | Anti-Dumping (Article VI of GATT 1994): Art. 2, 3, 5, 6, 7.1 GATT 1994: Art. VI, X | On 1 July 1996, Mexico requested consultations with the United States regarding the anti-dumping investigation on fresh and chilled tomatoes imported from Mexico under Article 17.3 of the Anti-Dumping Agreement. Violations of GATT Articles VI and X as well as Articles 2, 3, 5, 6 and 7.1 of the Anti-Dumping Agreement are alleged. Mexico claims this to be a case of urgency, where the expedited procedures under Articles 4.8 and 4.9 of the DSU are applicable. |

*(Continued)*

| Resolution year | Case | Dispute date and status | Parties | | Agreements cited | Summary |
|---|---|---|---|---|---|---|
| | | | Complainant | Respondent | | |
| 2001 | DISPUTE DS58 United States—Import Prohibition of Certain Shrimp and Shrimp Products | Compliance proceedings completed without finding of non-compliance on 21 November 2001 (up-to-date at 24 February 2010) | India; Malaysia; Pakistan; Thailand | United States **Third parties:** Australia; Canada; Colombia; Costa Rica; European Communities; Ecuador; El Salvador; Guatemala; Hong Kong, China; Japan; Mexico; Nigeria; Philippines; Senegal; Singapore; Sri Lanka; Venezuela, Bolivarian Republic of; Pakistan; Thailand | GATT 1994: Art. I, XI, XIII, XX | **Measure at issue:** US import prohibition of shrimp and shrimp products from non-certified countries (*i.e.,* countries that had not used a certain net in catching shrimp). **Product at issue:** Shrimp and shrimp products from the complainant countries. |

*(Continued)*

| Resolution year | Case | Parties | | Agreements cited | Summary |
| --- | --- | --- | --- | --- | --- |
| | | Complainant | Respondent | | |
| | | | | | On the grounds that the United States had not implemented appropriately the recommendations of the DSB, on 12 October 2000, Malaysia requested that the matter be referred to the original panel pursuant to Article 21.5 of the DSU. In particular, Malaysia considered that by not lifting the import prohibition and not taking the necessary measures to allow the importation of certain shrimp and shrimp products in an unrestrictive manner, the United States had failed to comply with the recommendations and rulings of the DSB. At its meeting of 23 October 2000, the DSB referred the matter to the original panel pursuant to Article 21.5 DSU. Australia, Canada, the European Communities, Ecuador, India, Japan, Mexico, Pakistan, Thailand, and Hong Kong, China reserved their third-party rights. On 8 November 2000, the compliance panel was composed. |

*(Continued)*

| Resolution year | Case | Dispute date and status | Parties | | Agreements cited | Summary |
| --- | --- | --- | --- | --- | --- | --- |
| | | | Complainant | Respondent | | |
| | | | | | | • The compliance panel circulated its report on 15 June 2001. The compliance panel concluded that:<br>• The measure adopted by the United States in order to comply with the recommendations and rulings of the DSB violated Article XI:1 of the GATT 1994.<br>• In light of the recommendations and rulings of the DSB, Section 609 of Public Law 101-162, as implemented by the Revised Guidelines of 8 July 1999 and as applied so far by US authorities, was justified under Article XX of the GATT 1994 as long as the conditions stated in the findings of this Report, in particular the ongoing serious good faith efforts to reach a multilateral agreement, remain satisfied. |

*(Continued)*

| Resolution year | Case | Dispute date and status | Parties | | Agreements cited | Summary |
|---|---|---|---|---|---|---|
| | | | Complainant | Respondent | | |
| | | | | | | Should any one of the conditions referred to above cease to be met in the future, the recommendations of the DSB may no longer be complied with. In such a case, any complaining party in the original case may be entitled to have further recourse to Article 21.5 of the DSU. See http://www.wto.org/ english/tratop_e/dispu_e/ cases_e/1pagesum_e/ ds58sum_e.pdf |
| 1996 | DISPUTE DS61 United States—Import Prohibition of Certain Shrimp and Shrimp Products | In consultations on 25 October 1996 (up-to-date at 24 February 2010) | Philippines | United States | GATT 1994: Art. I, II, III, VIII, XI, XIII Technical Barriers to Trade (TBT): Art. 2 | On 25 October 1996, the Philippines requested consultations with the United States in respect of a complaint by the Philippines regarding a ban on the importation of certain shrimp and shrimp products from the Philippines imposed by the United States under Section 609 of United States Public Law 101-62. Violations of Articles I, II, III, VIII, XI and XIII of GATT 1994 and Article 2 of the TBT Agreement are alleged. A nullification and impairment of benefits under GATT 1994 is also alleged. (See WT/DS58). |

*(Continued)*

| Resolution year | Case | Dispute date and status | Parties | | Agreements cited | Summary |
|---|---|---|---|---|---|---|
| | | | Complainant | Respondent | | |
| 1997 | DISPUTE DS66 Japan—Measures Affecting Imports of Pork | In consultations on 15 January 1997 (up-to-date at 24 February 2010) | EU | Japan | GATT 1994: Art. I, X, XIII | Complaint by the European Communities. On 15 January 1997, the EC requested consultations with Japan in respect of certain measures affecting imports of pork and its processed products imposed by Japan. The EC contended that these measures are in violation of Japan's obligations under Articles I, X:3 and XIII of the GATT 1994. The EC also contended that these measures nullify or impair benefits accruing to it under the GATT 1994. |
| 1998 | DISPUTE DS69 European Communities—Measures Affecting Importation of Certain Poultry Products | Report(s) adopted, with recommendation to bring measure(s) into conformity on 23 July 1998 (up-to-date at 24 February 2010) | Brazil | EU **Third parties:** Thailand; United States | Agriculture: Art. 4, 5 GATT 1994: Art. II, III, X, XIII, XXVIII Import Licensing: Art. 1, 3 | On 24 February 1997, Brazil requested consultations with the EC in respect of the EC regime for the importation of certain poultry products and the implementation by the EC of the Tariff Rate Quota for these products. Brazil contended that the EC measures are inconsistent with Articles X and XXVII of GATT 1994 and Articles 1 and 3 of the Agreement on Import Licensing Procedures. Brazil also contended that the measures nullify or impair benefits accruing to it directly or indirectly under GATT 1994. *(Continued)* |

| Resolution year | Case | Dispute date and status | Parties | | Agreements cited | Summary |
|---|---|---|---|---|---|---|
| | | | Complainant | Respondent | | |
| | DISPUTE DS72 European Communities—Measures Affecting Butter Products | Settled or terminated (withdrawn, mutually agreed solution) on 11 November 1999 (up-to-date at 24 February 2010) | New Zealand | EU **Third party:** United States | GATT 1994: Art. II, III, X, X:1, XI, XI:1, II:1(b) Import Licensing: Art. 3, 3.3 Technical Barriers to Trade (TBT): Art. 2, 2.2, 2.9, 2.11, 2.12 | This request, dated 24 March 1997, is in respect of decisions by the EC and the United Kingdom's Customs and Excise Department, to the effect that New Zealand butter manufactured by the ANMIX butter-making process and the spreadable butter-making process be classified so as to be excluded from eligibility for New Zealand's country-specific tariff quota established by the European Communities' WTO Schedule. New Zealand alleges violations of Articles II, X and XI of GATT, Article 2 of the TBT Agreement, and Article 3 of the Agreement on Import Licensing Procedures. This request, dated 24 March 1997, is in respect of decisions by the EC and the United Kingdom's Customs and Excise Department, to the effect that New Zealand butter manufactured by the ANMIX butter-making process and the spreadable butter-making process be classified so as to be excluded from eligibility for New Zealand's country-specific tariff quota established by the European Communities' WTO Schedule. New Zealand alleges violations of Articles II, X and XI of GATT, Article 2 of the TBT Agreement, and Article 3 of the Agreement on Import Licensing Procedures. |

*(Continued)*

| Resolution year | Case | Dispute date and status | Parties Complainant | Parties Respondent | Agreements cited | Summary |
|---|---|---|---|---|---|---|
| 1998 | DISPUTE DS74 Philippines—Measures Affecting Pork and Poultry | Settled or terminated (withdrawn, mutually agreed solution) on 13 January 1998 (up-to-date at 24 February 2010) | United States | Philippines | Agriculture: Art. 4 GATT 1994: Art. III, X, XI Import Licensing: Art. 1, 3 Trade-Related Investment Measures (TRIMs): Art. 2, 5 | This request, dated 1 April 1997, is in respect of the implementation by the Philippines of its tariff-rate quotas for pork and poultry. The United States contends that the Philippines' implementation of these tariff-rate quotas, in particular the delays in permitting access to the in-quota quantities and the licensing system used to administer access to the in-quota quantities, appears to be inconsistent with the obligations of the Philippines under Articles III, X, and XI of GATT 1994, Article 4 of the Agreement on Agriculture, Articles 1 and 3 of the Agreement on Import Licensing Procedures, and Articles 2 and 5 of TRIMs. The United States further contends that these measures appear to nullify or impair benefits accruing to it directly or indirectly under cited agreements. |

*(Continued)*

| Resolution year | Case | Dispute date and status | Parties | | Agreements cited | Summary |
|---|---|---|---|---|---|---|
| | | | Complainant | Respondent | | |
| 1997 | DISPUTE DS97 United States—Countervailing Duty Investigation of Imports of Salmon from Chile | In consultations on 5 August 1997 (up-to-date at 24 February 2010) | Chile | United States | Subsidies and Countervailing Measures: Art. 11 | On 5 August 1997, Chile requested consultations with the United States in respect of a countervailing duty investigation initiated by the US Department of Commerce against imports of salmon from Chile. Chile contended that the decision to initiate an investigation was taken in the absence of sufficient evidence of injury, in violation of Article 11.2 and 11.3. Chile also contended a violation of Article 11.4, in relation to the representative status of producers of salmon fillets. |
| 2000 | DISPUTE DS98 Korea—Definitive Safeguard Measure on Imports of Certain Dairy Products | Implementation notified by respondent on 26 September 2000 (up-to-date at 24 February 2010) | EU | Korea, Republic of **Third party:** Korea, Republic of | GATT 1994: Art. XIX Safeguards: Art. 2, 4, 5, 12 | **Measure at issue:** Definitive safeguard measure. **Product at issue:** Imports of certain dairy products (skimmed milk powder preparations). |

*(Continued)*

| Resolution year | Case | Parties | | Agreements cited | Summary |
| --- | --- | --- | --- | --- | --- |
| | | Complainant | Respondent | | |
| | | | | | GATT Art. XIX:1(a) (unforeseen developments): Reversing the Panel's legal reasoning, the Appellate Body held that the clause—"as a result of unforeseen development and of the effect of the obligations incurred by a contracting party under this Agreement, including tariff concessions"—in Art. XIX:1(a), although not an independent condition, describes certain circumstances which must be demonstrated as a matter of fact in order for a safeguard measure to be applied consistently with the requirements of Art. XIX. The Appellate Body concluded that the phrase "as a result of unforeseen developments" requires that the developments that led to a product being imported in such quantities and under such conditions as to cause or threaten to cause serious injury to domestic producers must have been "unexpected." The Appellate Body could not complete the Panel's analysis, however, due to the lack of undisputed facts in the record. |

*(Continued)*

| Resolution year | Case | Dispute date and status | Parties | | Agreements cited | Summary |
|---|---|---|---|---|---|---|
| | | | Complainant | Respondent | | |
| | | | | | | SA Art. 4.2(a) and (c) (injury determination—serious injury): The Appellate Body upheld the Panel's finding that Korea's serious injury determination did not meet the requirements of Art. 4.2, as it did not adequately examine all serious injury factors listed in Art. 4.2 (e.g., imports increase, market share, sales, production, productivity, etc.) and neither did it provide sufficient reasoning in its explanations of how certain factors support, or detract from, a finding of serious injury.<br>*(Continued)* |

| Resolution year | Case | Dispute date and status | Parties | | Agreements cited | Summary |
| | | | Complainant | Respondent | | |
| --- | --- | --- | --- | --- | --- | --- |
| 1997 | DISPUTE DS100 United States—Measures Affecting Imports of Poultry Products | In consultations on 18 August 1997 (up-to-date at 24 February 2010) | EU | United States | GATT 1994: Art. I, III, X, XI Sanitary and Phytosanitary Measures (SPS): Art. 2, 3, 4, 5, 8 Technical Barriers to Trade (TBT): Art. 2, 5 | On 18 August 1997, the EC requested consultations with the United States in respect of a ban on imports of poultry and poultry products from the EC by the US Department of Agriculture's Food Safety Inspection Service, and any related measures. The EC contended that although the ban is allegedly on grounds of product safety, the ban does not indicate the grounds upon which EC poultry products have suddenly become ineligible for entry into the US market. The EC considered that the ban is inconsistent with Articles I, III, X and XI of GATT 1994, Articles 2, 3, 4, 5, 8, and Annex C of the SPS Agreement, or Article 2 and 5 of the TBT Agreement. *(Continued)* |

| Resolution year | Case | Dispute date and status | Parties | | Agreements cited | Summary |
| | | | Complainant | Respondent | | |
| --- | --- | --- | --- | --- | --- | --- |
| 1997 | DISPUTE DS101 Mexico—Anti-Dumping Investigation of High-Fructose Corn Syrup (HFCS) from the United States | In consultations on 4 September 1997 (up-to-date at 24 February 2010) | United States | Mexico | Anti-dumping (Article VI of GATT 1994): Art. 2, 4, 5, 6 | On 4 September 1997, the United States requested consultations with Mexico in respect of an anti-dumping investigation of high-fructose corn syrup (HFCS) from the United States conducted by Mexico, resulting in a preliminary determination of dumping and injury, and the consequent imposition of provisional measures on imports of HFCS from the United States. The United States alleged violations of Articles 5.5, 6.1.3, 6.2, 6.4, and 6.5 of the Anti-Dumping Agreement. On 8 May 1998, the United States requested consultations in respect of the same anti-dumping investigation, which had resulted in the imposition of definitive anti-dumping measures on these imports from the United States. See WT/DS132 and WT/DS132/RW. *(Continued)* |

| Resolution year | Case | Dispute date and status | Parties | | Agreements cited | Summary |
|---|---|---|---|---|---|---|
| | | | Complainant | Respondent | | |
| 1998 | DISPUTE DS102 Philippines—Measures Affecting Pork and Poultry | Settled or terminated (withdrawn, mutually agreed solution) on 13 January 1998 (up-to-date at 24 February 2010) | United States | Philippines | Agriculture: Art. 4 GATT 1994: Art. III, X, XI Import Licensing: Art. 1, 3 Trade-Related Investment Measures (TRIMs): Art. 2, 5 | This request, dated 7 October 1997, is in respect of the same measures complained of by the United States in WT/DS74, but also includes Administrative Order No. 8, Series of 1997, which purports to amend the original measure complained of in WT/DS74. |
| 2003 | DISPUTE DS103 Canada—Measures Affecting the Importation of Milk and the Exportation of Dairy Products | Mutually acceptable solution on implementation notified on 9 May 2003 (up-to-date at 24 February 2010) | United States | Canada **Third parties:** Argentina; Australia; European Communities; Japan; Mexico | Agriculture: Art. 3, 4, 8, 9, 10 GATT 1994: Art. X, XI, XIII Import Licensing: Art. 2 Subsidies and Countervailing Measures: Art. 3 | **Measure at issue:** Canadian government's support system (Special Milk Classes Scheme) for domestic milk production and export, as well as Canada's tariff rate quota ("TRQ") regime for imports of fluid milk. **Industry at issue:** Milk and dairy product industry. See http://www.wto.org/english/tratop_e/dispu_e/cases_e/1pagesum_e/ds103sum_e.pdf |

(*Continued*)

| Resolution year | Case | Dispute date and status | Parties | | Agreements cited | Summary |
|---|---|---|---|---|---|---|
| | | | Complainant | Respondent | | |
| 1997 | DISPUTE DS104 European Communities—Measures Affecting the Exportation of Processed Cheese | In consultations on 23 October 1997 (up-to-date at 24 February 2010) | United States | EU | Agriculture: Art. 8, 9, 10, 11 Subsidies and Countervailing Measures: Art. 3 | On 8 October 1997, the United States requested consultations with the EC in respect of export subsidies allegedly granted by the EC on processed cheese without regard to the export subsidy reduction commitments of the EC. The United States contended that these measures by the EC distort markets for dairy products and adversely affect US sales of dairy products. The United States alleged violations of Articles 8, 9, 10, and 11 of the Agreement on Agriculture, and Article 3 of the Subsidies Agreement. *(Continued)* |

| Resolution year | Case | Dispute date and status | Parties | | Agreements cited | Summary |
|---|---|---|---|---|---|---|
| | | | Complainant | Respondent | | |
| 2012 | DISPUTE DS105 European Communities—Regime for the Importation, Sale and Distribution of Bananas | Settled or terminated (withdrawn, mutually agreed solution) on 8 November 2012 (up-to-date at 20 November 2012) | Panama | EU | Services (GATS): Art. II, XVII GATT 1994: Art. I, II, XIII Import Licensing: Art. 1, 3 Trade-Related Investment Measures (TRIMs): Art. 2, 5 | On 24 October 1997, Panama requested consultations with the EC in respect of the EC's regime for the importation, sale, and distribution of bananas as established through Regulation 404/93, as well as any subsequent legislation, regulations, or administrative measures adopted by the EC, including those reflecting the Framework Agreement on Bananas. Panama did not specify the WTO provisions which the EC regime violates. This is the same regime that was the subject of a successful challenge by the United States, Ecuador, Guatemala, Honduras, and Mexico (WT/DS27). |

*(Continued)*

| Resolution year | Case | Dispute date and status | Parties | | Agreements cited | Summary |
|---|---|---|---|---|---|---|
| | | | Complainant | Respondent | | |
| 1997 | DISPUTE DS111 United States—Tariff Rate Quota for Imports of Groundnuts | In consultations on 19 December 1997 (up-to-date at 24 February 2010) | Argentina | United States | Agriculture: Art. 1, 4, 15 GATT 1994: Art. II, X, XIII Import Licensing: Art. 1 Rules of Origin: Art. 2 | On 19 December 1997, Argentina requested consultations with the United States with respect to the alleged commercial detriment to Argentina resulting from a restrictive interpretation by the United States of the tariff rate quota negotiated by the two countries during the Uruguay Round, regarding the importation of groundnuts. Argentina alleged violations of Articles II, X, and XII of GATT 1994, Articles 1, 4, and 15 of the Agreement on Agriculture, Article 2 of the Agreement on Rules of Origin, and Article 1 of the Import Licensing Agreement. Nullification and impairment of benefits is also alleged. *(Continued)* |

| Resolution year | Case | Dispute date and status | Parties | | Agreements cited | Summary |
| --- | --- | --- | --- | --- | --- | --- |
| | | | Complainant | Respondent | | |
| 2003 | DISPUTE DS113 Canada—Measures Affecting Dairy Exports | Mutually acceptable solution on implementation notified on 9 May 2003 (up-to-date at 24 February 2010) | New Zealand | Canada **Third parties:** Argentina; Australia; European Communities; Japan; Mexico; United States | Agriculture: Art. 3, 8, 9, 10 GATT 1994: Art. X:1 | On 29 December 1997, New Zealand requested consultations with Canada in respect of an alleged dairy export subsidy scheme commonly referred to as the "special milk classes" scheme. New Zealand contended that the Canadian "special milk classes" scheme is inconsistent with Article XI of the GATT 1994, and Articles 3, 8, 9 and 10 of the Agreement on Agriculture. See http://www.wto.org/ english/tratop_e/dispu_e/ cases_e/1pagesum_e/ ds113sum_e.pdf |

*(Continued)*

| Resolution year | Case | Dispute date and status | Parties | | Agreements cited | Summary |
| | | | Complainant | Respondent | | |
|---|---|---|---|---|---|---|
| 2001 | DISPUTE DS132 Mexico—Anti-Dumping Investigation of High-Fructose Corn Syrup (HFCS) from the United States | Compliance proceedings completed with finding(s) of non-compliance on 21 November 2001 (up-to-date at 24 February 2010) | United States | Mexico **Third parties:** Jamaica; Mauritius; European Union | Anti-dumping (Article VI of GATT 1994): Art. 1, 2, 3, 4, 5, 6, 7, 7.4, 9, 10, 10.2, 10.4, 12 | On 8 May 1998, the United States requested consultations with Mexico in respect of an anti-dumping investigation of high-fructose corn syrup (HFCS) grades 42 and 55 from the United States, conducted by Mexico. The United States alleged that on 27 February 1997, the Government of Mexico published a notice initiating this anti-dumping investigation on the basis of an application dated 14 January 1997 from the Mexican National Chamber of Sugar and Alcohol Producers. The United States further alleged that on 23 January 1998, Mexico issued a notice of final determination of dumping and injury in that investigation, and consequently imposed definitive anti-dumping measures on these imports from the United States. The United States contended that the manner in which the application for an anti-dumping investigation was made, as well as the manner in which a determination of threat of injury was made, is inconsistent with Articles 2, 3, 4, 5, 6, 7, 9, 10, and 12 of the Anti-Dumping Agreement. |

*(Continued)*

| Resolution year | Case | Dispute date and status | Parties | | Agreements cited | Summary |
| | | | Complainant | Respondent | | |
| --- | --- | --- | --- | --- | --- | --- |
| | | | | | | See http://www.wto.org/english/tratop_e/dispu_e/cases_e/1pagesum_e/ds132sum_e.pdf |
| 1998 | DISPUTE DS133 Slovak Republic—Measures Concerning the Importation of Dairy Products and the Transit of Cattle | In consultations on 7 May 1998 (up-to-date at 24 February 2010) | Switzerland | Slovak Republic | GATT 1994: Art. I, III, V, X, XI Import Licensing: Art. 5 Sanitary and Phytosanitary Measures (SPS): Art. 5 | On 7 May 1998, Switzerland requested consultations with the Slovak Republic concerning measures imposed by the Slovak Republic (in particular, a decree of 6 July 1996) with respect to the importation of dairy products and the transit of cattle. Switzerland contended that these measures had a negative impact on Swiss exports of cheese and cattle. Switzerland alleged that some of these measures are inconsistent with Articles I, III, V, X, and XI of GATT 1994, Article 5 of the SPS Agreement, and Article 5 of the Import Licensing Agreement. |

*(Continued)*

| Resolution year | Case | Dispute date and status | Parties | | Agreements cited | Summary |
|---|---|---|---|---|---|---|
| | | | Complainant | Respondent | | |
| 1998 | DISPUTE DS134 European Communities—Restrictions on Certain Import Duties on Rice | In consultations on 27 May 1998 (up-to-date at 24 February 2010) | India | EU | Agriculture: Art. 4 GATT 1994: Art. I, II, III, VIII, XI Import Licensing: Art. 1, 3 Sanitary and Phytosanitary Measures (SPS): Art. 2 Technical Barriers to Trade (TBT): Art. 2 Customs valuation (Article VII of GATT 1994): Art. 1, 2, 3, 4, 5, 6, 7, 11 | On 27 May 1998, India requested consultations with the EC in respect of the restrictions allegedly introduced by an EC Regulation establishing a so-called cumulative recovery system (CRS), for determining certain import duties on rice, with effect from 1 July 1997. India contended that the measures introduced through this new regulation will restrict the number of importers of rice from India, and will have a limiting effect on the export of rice from India to the EC. India alleged violations of Articles I, II, III, VII, and XI of GATT 1994, Articles 1–7, 11 and Annex I of the Customs Valuation Agreement, Articles 1 and 3 of the Import Licensing Agreement, Article 2 of the TBT Agreement, Article 2 of the SPS Agreement, and Article 4 of the Agreement on Agriculture. India also claimed nullification and impairment of benefits accruing to it under the various agreements cited. |

*(Continued)*

| Resolution year | Case | Dispute date and status | Parties | | Agreements cited | Summary |
|---|---|---|---|---|---|---|
| | | | Complainant | Respondent | | |
| 1998 | DISPUTE DS137 European Communities—Measures Affecting Imports of Wood of Conifers from Canada | In consultations on 17 June 1998 (up-to-date at 24 February 2010) | Canada | EU | GATT 1994: Art. I, III, XI Sanitary and Phytosanitary Measures (SPS): Art. 2, 3, 4, 5, 6 Technical Barriers to Trade (TBT): Art. 2 | On 17 June 1998, Canada requested consultations with the EC in respect of certain measures concerning the importation into the EC market of wood of conifers from Canada. The measures include, but are not limited to, Council Directive 77/93, of 21 December 1976, as amended by Commission Directive 92/103/EEC, of 1 December 1992, and any relevant measures adopted by EC Member states affecting imports of wood of conifers from Canada into the EC. Canada alleged that these adversely affect the importation into the EC market of wood of conifers from Canada. Canada alleged violations of Articles I, III, and XI of GATT 1994, Articles 2, 3, 4, 5, and 6 of the SPS Agreement, and Article 2 of the TBT Agreement. Canada also made a claim for nullification and impairment of benefits accruing to it indirectly under the cited agreements. |

*(Continued)*

| Resolution year | Case | Parties | | Dispute date and status | Agreements cited | Summary |
|---|---|---|---|---|---|---|
| | | Complainant | Respondent | | | |
| 1998 | DISPUTE DS143 Slovak Republic—Measure Affecting Import Duty on Wheat from Hungary | Hungary | Slovak Republic | In consultations on 8 October 1998 (up-to-date at 12 October 2012) | Agriculture: Art. 4 GATT 1994: Art. I, II | On 18 September 1998, Hungary requested consultations with the Slovak Republic in respect of a regulation adopted by the Slovak Republic which entered into force on 10 September 1998, which allegedly increased the import duty of wheat originating in Hungary. Hungary asserted that the increased import duty on wheat (HS1001.1000, 1001.90) amounts to 2,540 SKK/t which equals to approximately 70% ad valorem. Hungary alleged that: the bound rates for these tariff lines in the Slovak Schedule for 1998 are set at 4.4% (HS1001.1000), 27% (HS1001.9010) and 22.5% (HS1001.9091, 1001.9099); it is the only country subject to this measure; and this measure is inconsistent with Articles I and II of GATT 1994, and Article 4 of the Agreement on Agriculture. Hungary invoked the urgency provision of the DSU due to the severe economic and trade losses that are being caused by this measure, which was expected to remain in force until 10 March 1999. |

*(Continued)*

| Resolution year | Case | Dispute date and status | Parties | | Agreements cited | Summary |
|---|---|---|---|---|---|---|
| | | | Complainant | Respondent | | |
| 1998 | DISPUTE DS144 United States—Certain Measures Affecting the Import of Cattle, Swine, and Grain from Canada | In consultations on 25 September 1998 (up-to-date at 24 February 2010) | Canada | United States | Agriculture: Art. 4 GATT 1994: Art. I, III, V, XI, XXIV:12 Sanitary and Phytosanitary Measures (SPS): Art. 2, 3, 4, 5, 6, 13, Annex B, Annex C Technical Barriers to Trade (TBT): Art. 2, 3, 5, 7 | On 25 September 1998, Canada requested consultations with the United States in respect of certain measures, imposed by the US state of South Dakota and other states, prohibiting entry or transit to Canadian trucks carrying cattle, swine, and grain. Canada alleged that these measures adversely affect the importation into the United States of cattle, swine, and grain originating in Canada. Canada alleges violations of Articles 2, 3, 4, 5, 6, 13, and Annexes B and C of the SPS Agreement; Articles 2, 3, 5, and 7 of the TBT Agreement; Article 4 of the Agreement on Agriculture; and Articles I, III, V, XI, and XXIV:12 of GATT 1994. Canada also made a claim of nullification or impairment of benefits accruing to it under the cited Agreements. Canada invoked Article 4.8 of the DSU for expedited consultations in view of the perishable nature of the goods in question. |

*(Continued)*

| Resolution year | Case | Dispute date and status | Parties | | Agreements cited | Summary |
|---|---|---|---|---|---|---|
| | | | Complainant | Respondent | | |
| 1998 | DISPUTE DS145 Argentina—Countervailing Duties on Imports of Wheat Gluten from the European Communities | In consultations on 7 October 1998 | EU | Argentina | Subsidies and Countervailing Measures: Art. 10, 11.11 | On 23 September 1998, the EC requested consultations with the Argentina in respect of definitive countervailing duties allegedly imposed by Argentina on imports of wheat gluten from the EC. The EC stated that Argentina imposed a countervailing duty on wheat gluten imports from the EC with effect from 23 July 1998. The investigation which led to the imposition of these duties had been initiated on 23 October 1996 and, consequently, the EC contended that the investigation exceeded 18 months, contrary to Article 11.11 of the Subsidies Agreement. The EC also claimed a violation of Article 10 of the same Agreement. *(Continued)* |

| Resolution year | Case | Dispute date and status | Parties | | Agreements cited | Summary |
|---|---|---|---|---|---|---|
| | | | Complainant | Respondent | | |
| 1998 | DISPUTE DS148 Czech Republic—Measure Affecting Import Duty on Wheat from Hungary | In consultations on 12 October 1998 (up-to-date at 24 February 2010) | Hungary | Czech Republic | Agriculture: Art. 4 GATT 1994: Art. I, II | On 12 October 1998, Hungary requested consultations with the Czech Republic in respect of a regulation adopted by the Czech Republic which entered into force on 9 October 1998, and which allegedly increased the import duty of wheat originating in Hungary. Hungary asserted that the increased import duty on wheat (HS1001.1000, 1001.9099) exceeds several times the respective bound rates in the Czech Schedule for 1998. Hungary also alleged that it is the only country subject to this measure. Hungary contended that this measure is inconsistent with Articles I and II of GATT 1994, and Article 4 of the Agreement on Agriculture. Hungary invoked the urgency provision of the DSU (4.8), due to the severe economic and trade losses that are being caused by this measure, which was expected to remain in force until 26 April 1999. |

*(Continued)*

| Resolution year | Case | Dispute date and status | Parties | | Agreements cited | Summary |
|---|---|---|---|---|---|---|
| | | | Complainant | Respondent | | |
| 1998 | DISPUTE DS154 European Communities—Measures Affecting Differential and Favorable Treatment of Coffee | In consultations on 7 December 1998 (up-to-date at 24 February 2010) | Brazil | EU | GATT 1994: Art. I | On 7 December 1998, Brazil requested consultations with the EC in respect of the special preferential treatment under the EC's Generalised System of Preferences (GSP). Brazil asserted that the EC GSP scheme is applicable to products originating in the Andean Group of countries and the Central American Common Market countries, that are conducting programs to combat drug production and trafficking. In the case of soluble coffee, this special preferential treatment, contained in Council Regulation (EC) No. 1256/96, amounts to duty free access into the EC market. Brazil stated that it is aware that there is a proposed Council Regulation which would unify all EC laws and regulations concerning the operation of the GSP scheme for both agricultural and industrial products. Brazil contended that this special treatment adversely affects the importation into the EC of soluble coffee originating in Brazil. Brazil alleged that this special treatment is inconsistent with the Enabling Clause, as well as with Article I of GATT 1994. Brazil further alleges that this special treatment nullifies or impairs benefits accruing to Brazil directly or indirectly under the cited provisions. |

*(Continued)*

| Resolution year | Case | Dispute date and status | Parties | | Agreements cited | Summary |
| | | | Complainant | Respondent | | |
| --- | --- | --- | --- | --- | --- | --- |
| 2012 | DISPUTE DS158 European Communities—Regime for the Importation, Sale and Distribution of Bananas | Settled or terminated (withdrawn, mutually agreed solution) on 8 November 2012 (up-to-date at 20 November 2012) | Guatemala; Honduras; Mexico; Panama; United States | EU | Import Licensing: Art. 6 | On 20 January 1999, these countries (complaining parties) requested consultations with the EC in respect of the implementation of the recommendations of the DSB in European Communities—Regime for the Importation, Sale and Distribution of Bananas. The complaining parties state that the 15-month reasonable period of time for the EC to implement the DSB's recommendations and rulings ended on 1 January 1999 (see WT/DS27). The complaining parties alleged that the EC modified its regime in a manner that will not permit this dispute to conclude at this time on the basis of a solution that is acceptable to their governments, and as a result, jointly and severally, request consultations with the EC concerning the EC banana regime established by EC Regulation 404/93, as amended and implemented by Council Regulation 1637/98 of 20 July 1998 and EC Commission Regulation 2362/98 of 28 October 1998. The complaining parties contended that their objective is to clarify and discuss in detail with the EC the various aspects of the EC's modified banana regime, including their effect on the market, their concerns about their WTO-inconsistency, and ways that the EC might modify its regime in order to produce a satisfactory settlement of this dispute. |

*(Continued)*

| Resolution year | Case | Dispute date and status | Parties | | Agreements cited | Summary |
|---|---|---|---|---|---|---|
| | | | Complainant | Respondent | | |
| 2001 | DISPUTE DS161 Korea—Measures Affecting Imports of Fresh, Chilled, and Frozen Beef | Implementation notified by respondent on 25 September 2001 (up-to-date at 24 February 2010) | United States | Korea, Republic of **Third parties:** Australia; Canada; New Zealand | Agriculture: Art. 3, 4, 6, 7 GATT 1994: Art. II, III, X, XI, XVII Import Licensing: Art. 1, 3 | On 1 February 1999, the United States requested consultations with Korea in respect of a Korean regulatory scheme that allegedly discriminates against imported beef by *inter alia*, confining sales of imported beef to specialized stores (dual retail system), limiting the manner of its display, and otherwise constraining the opportunities for the sale of imported beef. The United States alleged that Korea imposes a mark-up on sales of imported beef, limits import authority to certain so-called "super-groups" and the Livestock Producers Marketing Organization ("LPMO"), and provides domestic support to the cattle industry in Korea in amounts which cause Korea to exceed its aggregate measure of support, as reflected in Korea's schedule. The United States contended that these restrictions apply only to imported beef, thereby denying national treatment to beef imports, and that the support to the domestic industry amounts to domestic subsidies that contravene the Agreement on Agriculture. The United States alleged violations of Articles II, III, XI, and XVII of GATT 1994; Articles 3, 4, 6, and 7 of the Agreement on Agriculture; and Articles 1 and 3 of the Import Licensing Agreement. *(Continued)* |

| Resolution year | Case | Dispute date and status | Parties | | Agreements cited | Summary |
|---|---|---|---|---|---|---|
| | | | Complainant | Respondent | | |
| 2001 | DISPUTE DS166 United States—Definitive Safeguard Measures on Imports of Wheat Gluten from the European Communities | Report(s) adopted, with recommendation to bring measure(s) into conformity on 19 January 2001 (up-to-date at 24 February 2010) | European Communities | United States **Third parties:** Australia; Canada; New Zealand | Agriculture: Art. 4.2 GATT 1994: Art. I, XIX Safeguards: Art. 2, 2.1, 4, 5, 8, 12 | **Measure at issue:** Definitive safeguard measure imposed by the United States. **Product at issue:** Wheat gluten from the European Communities. On 17 March 1999, the EC requested consultations with the United States in respect of definitive safeguard measures imposed by the United States on imports of wheat gluten from the European Communities. The EC contended that by a Proclamation of 30 May 1998, and a Memorandum of the same date, by the US President, under which the United States imposed definitive safeguard measures in the form of a quantitative limitation on imports of wheat gluten from the EC, effective as of 1 June 1998. The EC considered these measures to be in violation of Articles 2, 4, 5, and 12 of the Agreement on Safeguards; Article 4.2 of the Agreement on Agriculture; and Articles I and XIX of GATT 1994. |

*(Continued)*

| Resolution year | Case | Dispute date and status | Parties | | Agreements cited | Summary |
|---|---|---|---|---|---|---|
| | | | Complainant | Respondent | | |
| 1999 | DISPUTE DS167 United States—Countervailing Duty Investigation with respect to Live Cattle from Canada | In consultations on 19 March 1999 (up-to-date at 24 February 2010) | Canada | United States | Agriculture: Art. 13 Subsidies and Countervailing Measures: Art. 1, 2, 10, 11.1, 11.5, 13.1 | On 19 March 1999, Canada requested consultations with the United States concerning the initiation of a countervailing duty investigation by the United States, on 22 December 1998, with respect to live cattle from Canada. Canada alleged that: The initiation of this investigation is inconsistent with US obligations under the Subsidies Agreement, including the fact that the written application filed with the US Department of Commerce was not made by or on behalf of the domestic industry, and that there was not sufficient information provided with respect to the measures or actions alleged to be subsidies, for purpose of initiating an investigation under the SCM Agreement; the measures or actions alleged to be subsidies either are not, in law or fact, subsidies within the meaning of the Subsidies Agreement, or do not confer more than a *de minimis* level of countervailing subsidy; and this initiation of investigation is inconsistent with US obligations under the Agreement on Agriculture relating to "due restraint." |

*(Continued)*

| Resolution year | Case | Dispute date and status | Parties | | Agreements cited | Summary |
| | | | Complainant | Respondent | | |
| --- | --- | --- | --- | --- | --- | --- |
| 2001 | DISPUTE DS169 Korea—Measures Affecting Imports of Fresh, Chilled, and Frozen Beef | Implementation notified by respondent on 25 September 2001 (up-to-date at 24 February 2010) | Australia | Korea, Republic of **Third parties:** Canada; New Zealand; United States | Agriculture: Art. 3, 4, 6, 7 GATT 1994: Art. II, III, X, XI, XVI, XVII Import Licensing: Art. 1, 3 | On 1 February 1999, the United States requested consultations with Korea in respect of a Korean regulatory scheme that allegedly discriminates against imported beef by *inter alia*, confining sales of imported beef to specialized stores (dual retail system), limiting the manner of its display, and otherwise constraining the opportunities for the sale of imported beef. The United States alleged that Korea imposes a mark-up on sales of imported beef, limits import authority to certain so-called "super-groups" and the Livestock Producers Marketing Organization ("LPMO"), and provides domestic support to the cattle industry in Korea in amounts which cause Korea to exceed its aggregate measure of support as reflected in Korea's schedule. The United States contended that these restrictions apply only to imported beef, thereby denying national treatment to beef imports, and that the support to the domestic industry amounts to domestic subsidies that contravene the Agreement on Agriculture. |

*(Continued)*

| Resolution year | Case | Dispute date and status | Parties | | Agreements cited | Summary |
|---|---|---|---|---|---|---|
| | | | Complainant | Respondent | | |
| 2001 | DISPUTE DS177 United States—Safeguard Measure on Imports of Fresh, Chilled or Frozen Lamb from New Zealand | Implementation notified by respondent on 21 November 2001 (up-to-date at 24 February 2010) | New Zealand | United States **Third parties:** Australia; Canada; European Communities; Iceland; Japan | GATT 1994: Art. I, II, XIX Safeguards: Art. 2, 3, 4, 5, 11, 12 | On 16 July 1999, New Zealand requested consultations with the United States in respect of a safeguard measure imposed by the United States on imports of lamb meat from New Zealand (WT/DS177). New Zealand alleged that by Presidential Proclamation under Section 203 of the US Trade Act 1974, the United States imposed a definitive safeguard measure in the form of a tariff-rate quota on imports fresh, chilled, or frozen lamb meat effective from 22 July 1999. New Zealand contended that this measure is inconsistent with Articles 2, 4, 5, 11 and 12 of the Agreement on Safeguards, and Articles I and XIX of GATT 1994. |

*(Continued)*

| Resolution year | Case | Dispute date and status | Parties | | Agreements cited | Summary |
| --- | --- | --- | --- | --- | --- | --- |
| | | | Complainant | Respondent | | |
| | | | | | | On 23 July 1999, Australia requested consultations with the United States in respect of a definitive safeguard measure imposed by the United States on imports of lamb (WT/DS178). Australia alleged that by Presidential Proclamation under Section 203 of the US Trade Act 1974, the United States imposed a definitive safeguard measure in the form of a tariff-rate quota on imports of fresh, chilled, or frozen lamb meat from Australia effective from 22 July 1999. Australia contended that this measure is inconsistent with Articles 2, 3, 4, 5, 8, 11, and 12 of the Agreement on Safeguards, and Articles I, II, and XIX of GATT 1994. |

## chapter seven

# Concluding remarks:
# Obstacles to food integrity

The commonality of Chapters 1 through 6 rests in the identification of systematic flaws in our food system, both domestic and international, that enables the trivialization of genetically modified organisms (GMOs) and their proliferation, which undermine food safety, security, and sovereignty. Using Figure 2.1 as an anchor for the perspective shifts and legal comparison in this book, the author has drawn several conclusions about the preservation of food integrity and obstacles thereto. This chapter first explains how communication challenges in food policy fuel the trivialization of GMO risks to the food system. Then, Sections 7.2 and 7.3 discuss sustainability and the polluter pays principles. Finally, the chapter concludes with an overarching conclusion about the limits of using the law to address the challenges of our food system.

## 7.1   Communication challenges in food policy:
##        Agencies, science, and uncertainty

Transparency in food labeling (Chapter 5), consumer access to trade measures (Chapter 6), and governmental sovereignty over the regulation and trade of GMOs essentially pose communication challenges. In US food law, for instance, advocates face communication challenges from the GMO risk regulation perspective because they face "standing challenges and justiciability issues under the Administrative Procedure Act"[1] and because consumers are sometimes not receptive or remain inactive citizens. Therefore, communication should be encouraged and spread more widely by mandating a greater amount of exposure and education. This would be accomplished by broadening and deepening food safety education and by putting some food safety control into consumers' hands.

First, achieving improved communication about food integrity requires that some degree of food safety education is part of every school's curriculum, so that children are sensitized to these issues and

---

[1] Diana Winters, Not sick yet: Food-safety-impact litigation and barriers to justiciability *Brooklyn Law Review*, Vol. 77, No. 3, p. 905, 2012, Boston Univ. School of Law, Public Law Research Paper No. 12-31.

grow up to become more receptive adults. Second, in current adults, one should "harness human self-interest to effect reform"[2] because "...agriculture generates many significant costs, including ecological, public health, economic, and social costs."[3] If people were aware of the public health costs of food safety disasters, they might realize that food safety regulation currently fails to (1) protect the integrity of the market; (2) regulate the nutritional content of food; and (3) protect the safety of the food supply.[4] With this realization, this author hopes that the general public will understand that the large food producers are often externalizing the costs of food production and shift much of the burden on the public for the producers' short-term profit. Next, consumers might become aware of agency capture and require more government oversight funding for consumer protection. Ultimately, through exposure to the facts and consequences of food safety oversight failures, the public will increase pressure on the government to resist the one-sided and short-sighted capitalization of GMOs.

Second, where the bottom-up demand for improved food safety oversight is too slow, education campaigns to put consumers in power through a supply-and-demand approach could be promising complements to using legal tools, such as regulations, embargoes, cases, and treaties. As such, consumers can take some measures of public control, for example, by supporting short supply chains from local foods that increase transparency and accountability. The US Department of Agriculture (USDA) even supports such measures through the "Know Your Farmer, Know Your Food" program[5] that incentivizes the supply-and-demand approach.

In conclusion, every consumer is a piece of the food safety regulatory and advocacy puzzle. Together, food safety, a core interest common to all people, is a powerful means of exerting pressure on the government through organized advocacy and from the public on a small scale. Armed with the proper facts and strategies, consumers can exert enormous pressure and may, hopefully, turn food safety inefficiencies around for the better.

---

[2] Elizabeth Hallinan and Jeffrey D. Pierce, Learning from patchwork environmental regulation: What animal advocates might learn from the varied history of the Clean Air Act, in *What Can Animal Law Learn From Environmental Law?* at 40 (Randall Abate, Ed.). Environmental Law Institute. Kindle edition.

[3] Id.

[4] Winters, *supra* note 2 at 910.

[5] USDA, Know Your Farmer, Know Your Food, https://www.usda.gov/sites/default/files/documents/KYFCompass.pdf.

## 7.2  Polluter pays: Insourcing of production externalities

What if existing legal tools were used to reverse the cycle depicted in Figure 1.2? For instance, could a pollution tariff partially solve the Chinese environmental problem where countries importing Chinese goods have outsourced industrial production and, thereby, pollution? Would this help the United States and the EU to insource its responsibility through a polluter pays model?

The pressure for continued economic growth and the demand for Chinese goods with an unsustainable environmental footprint may build upon the Chinese industry's economic drive and promote development toward environmental protection through a global pollution tariff. In an effort to internalize environmental costs on Chinese goods and exports, what would the legal or political obstacles be to a proposed international treaty to impose tariffs on goods that fail to conform to a certain set of higher pollution restrictions, such as those from the United States and the EU and as enforced by the World Trade Organization (WTO)? What would make such a treaty feasible and enforceable despite the WTO's current failure to intervene in environmental disputes among the major trading nations?

Two of the factors slowing down environmental protection in China seem to be (1) the stifled pressures from within the country due to the drive for economic growth, and (2) the ongoing global demand for goods that appear to be cheaply produced in China, but which come at great environmental cost. Thus, Japan, the United States, and the EU have externalized the environmental burden of industrially producing goods to China, a country with comparably inefficient environmental laws. While, in China, trade and environmental protection seem to be irreconcilable to a large degree due to internal economic pressures, tariffs may create outside pressures from economic growth to environmental protection.

A tariff would require "monitoring and verification, as well as a system to administer the funds created by the tariff." This author's hypothetical proposed pollution tariff could, for example, build on the Agreement on Technical Barriers to Trade (TBT), which is enforced by the WTO. The TBT, as summarized by the WTO,

> recognizes that countries have the right to establish protection, at levels they consider appropriate, for example for human, animal or plant life or health or the environment, and should not be prevented from taking measures necessary to ensure those levels of protection are met. The agreement therefore

encourages countries to use international standards
where these are appropriate, but it does not require
them to change their levels of protection as a result
of standardization.[6]

This means that, under the TBT, unjust enrichment of one country at
the expense of another should be struck down by the WTO. Pollution is
just an abstraction of this enrichment, not necessarily at the expense of
one trade partner, but at the expense of everyone bearing the cost of climate change and environmental degradation. Therefore, this author proposes that embracing the global limitations of resources by way of a tariff
might prevent future battles of resources as they run out, such as fresh
water and fossil fuels, so that growth may continue, albeit sustainable.

## 7.3   Coexistence paradox: GMOs at the exclusion of agroecology

It is about one or the other: the coexistence paradox of GMOs and agroecology cannot and will not work as long as GMOs are farmed at the expense
of agroecology and food integrity. As the previous six chapters described,
the private law protections of BigAg's proliferation of GMOs by trivializing the underlying risks undermine the efforts to farm in harmony with
ecologically sustainable processes.

## 7.4   Summary

This chapter summarized the commonalities in the identification of systematic flaws in our food system, both domestically and internationally,
that enable the trivialization of GMOs and their proliferation, which
undermine food safety, security, and sovereignty. Moreover, this chapter
summarized the book's conclusions about the preservation of food integrity and obstacles thereto. Finally, this chapter explained how communication challenges in food policy fuel the trivialization of GMO risks to the
food system and discusses sustainability and the polluter pays principles.
Finally, the chapter concluded with an overarching conclusion about the
limits of using the law to address the challenges of our food system.

---

[6] WTO, Legal texts: The WTO agreements: Agreement on technical barriers to trade,
https://www.wto.org/english/docs_e/legal_e/ursum_e.htm#dAgreement.

# Glossary

**Agricultural sustainability** shall be construed to complement environmental conservation and climate change mitigation.

**Agroecology** is the ecology of food systems.[1]

**Anthropogenic climate change** is defined as human-induced "widespread change detected in temperature observations of the surface, … free atmosphere … and ocean, together with consistent evidence of change in other parts of the climate system, … [which] strengthens the conclusion that greenhouse gas forcing is the dominant cause of warming during the past several decades."[2]

**Food dependence** describes people's reliance on, confidence in, and trust in food producers, contingent upon the farming practices of the industry and all of its components.

**Food integrity** is the measure of environmental sustainability and climate change resilience, combined with food safety, security, and sovereignty for the farm-to-fork production and distribution of any food product.[3]

**Food mile** denotes "a measure of the distance traveled by foods between the place where they are produced and the place where they are eaten. Long distances are considered bad because the quality of the food is worse, and energy is wasted in transporting it."[4]

---

[1] C. Francis, G. Lieblein, S. Gliessman, T.A. Breland, N. Creamer, R. Harwood, L. Salomonsson, J. Helenius, D. Rickerl, R. Salvador, M. Wiedenhoeft, S. Simmons, P. Allen, M. Altieri, C. Flora, and R. Poincelot, Agroecology: The ecology of food systems, *Journal of Sustainable Agriculture*, 22, 3 (2003), http://www.tandfonline.com/doi/abs/10.1300/J064v22n03_10 (last accessed May 29, 2017).

[2] IPCC, IPCC Fourth Assessment Report at 9.7. Combining evidence of anthropogenic climate change (2007), https://www.ipcc.ch/publications_and_data/ar4/wg1/en/ch9s9-7.html (last accessed May 30, 2017).

[3] G. Steier, A window of opportunity for GMO regulation: Achieving food integrity through cap-and-trade models from climate policy for GMO regulation, *Pace Envtl L. Rev.* 34, 293 (2017).

[4] *Macmillan Dictionary*, Food Mile, http://www.macmillandictionary.com/us/dictionary/american/food-mile (last accessed May 30, 2017).

**Food safety**  focuses on "handling, storing and preparing food to prevent infection and help to make sure that … food keeps enough nutrients for … a healthy diet."[5]

**Food security**  exists when "availability and adequate access at all times to sufficient, safe, nutritious food to maintain a healthy and active life"[6] are guaranteed. Three pillars of food security, (1) availability, (2) access, and (3) utilization, are designed to ensure that food is "available in sufficient quantities and on a consistent basis," that it is regularly acquirable at adequate quantities, and that it has "a positive nutritional impact on people."[7]

**Food sovereignty**  is the right of peoples to healthy and culturally appropriate food produced through ecologically sound and sustainable methods, and their right to define their own food and agriculture systems.[8]

**Sustainability**  shall be defined as pertaining to a food system that maintains its own viability by using agroecologic techniques that allow for continual reuse and holistic service to all components of food integrity.

---

[5] FAO, Food safety, http://www.fao.org/docrep/008/a0104e/a0104e08.htm.
[6] WFP, What is food security?, https://www.wfp.org/node/359289.
[7] Id.
[8] La Via Campesina, The Declaration of Nyéléni, https://nyeleni.org/spip.php?article290.

# *Index*

Printed and bound by CPI Group (UK) Ltd, Croydon, CR0 4YY

24/10/2024

01778304-0002